# International Mathematics

## Coursebook 1

Andrew Sherratt

HODDER EDUCATION
AN HACHETTE UK COMPANY

Hachette UK's policy is to use papers that are natural, renewable and recyclable products and mode from wood grown in well-managed forests and other controlled sources. The logging and manufacturing processes are expected to conform to the environmental regulations of the country of origin.

Orders: please contact Hachette UK Distribution, Hely Hutchinson Centre, Milton Road, Didcot, Oxfordshire, OX11 7HH. Telephone: +44 (0)1235 827827. Email education@hachette.co.uk. Lines are open from 9 a.m. to 5 p.m., Monday to Friday. You can also order through our website: www.hoddereducation.com

Impression number    12
Year            2022

Cover photo © Alan Schein Photography/Corbis
Illustrations by Barking Dog Art
Typeset in 12.5/15.5pt Garamond by Charon Tec Ltd (A Macmillan Company)

Printed and bound by CPI Group (UK) Ltd, Croydon, CR0 4YY

A catalogue record for this title is available from the British Library

ISBN 978 0 340 96742 3

# Contents

# Contents

# Introduction to Maths in English

These books are written specially for pupils who do not speak English at home, but are studying Maths in English at school.

I have used simple language and explanations that I hope will help you to understand the Maths without being confused by complicated English words and sentences.

Many pupils who do speak English at home will also find it useful to have a Maths book that explains things simply and clearly.

To help you, each unit starts with a list of important words (**Key vocabulary**) that you will need to understand before you start the work in that unit.

It is a very good idea to make your own 'Maths Wordbook'.

- Your teacher can show you how to divide an exercise book into sections, with a few pages for each letter of the alphabet.
- Each time you learn a new Maths word, write it in your wordbook. Keep all the words that start with the same letter together so that you know where to find them later on.
- Next to each word, write (or draw) something that will help you to remember what the word means, and how to use it.
- Do **not** copy what you find in a dictionary – this is a waste of time.
- Write (or draw) whatever will mean something to **you** when you look at it again. For example:
  - your own translation into your home language
  - a drawing or picture
  - an example of how to use the word.

To help you start your wordbook, on the next page is a list of some words that will help you to understand Maths in English. See how many of them you can learn before you begin Unit 1!

## Key vocabulary

| | | |
|---|---|---|
| above | increase | reduce |
| answer | increased by | repeat |
| basic | label | replace |
| below | larger | right |
| bigger than | larger than | right-hand side |
| calculate | largest | rise |
| calculation | left | rule |
| change | left-hand side less | separate |
| check | less than | separately |
| combine | lose | seven |
| cost | lower | six |
| count | measure | sign |
| decrease | missing | simplify |
| decreased by | more than | smaller than |
| direction | move | space |
| double | movement | statement |
| eight | negative | take away |
| equal to | negative whole number | third |
| equals | nine | three |
| exactly | none | times |
| example | one | together |
| false | opposite | total |
| five | pair | triple |
| four | plus | true |
| gain | problem | twice |
| greater | process | two |
| greater than | property | whole |
| greatest | quarter | work out |
| grow | question | |
| half | quick | |

Throughout each unit, there are many worked examples followed by exercises for you to practise and revise what you have learnt. You will notice that one of these icons sometimes appears next to an exercise:

 This means you **should** use a calculator to answer the questions.

 This means you should **not** use a calculator to answer the questions.

Good luck to you all – and I hope you enjoy studying Maths in English with this book!

Andrew Sherratt

# Unit 1 Integers

## Key vocabulary

| | | |
|---|---|---|
| add | integer | positive |
| addition | less | positive sign |
| ascending | long division | product |
| bigger | long multiplication | quotient |
| biggest | million | remainder |
| billion | minus | short division |
| borrow | more | short multiplication |
| bracket | multiplication | smaller |
| column | multiply | smallest |
| counting numbers | negative | subtract |
| descending | negative sign | subtraction |
| difference | number | sum |
| digit | number line | symbol |
| divide | numeral | tables |
| division | odd | thousand |
| equal | operation | unit |
| equation | order | value |
| even | parts | whole number |
| figure | place value | zero |

## A Positive whole numbers

The first group of numbers we learn about, when we are children, are the **whole numbers**. If something is **whole**, it is complete, with no **parts** missing or left over. Parts of numbers are called fractions or decimals, and we will learn about these in Units 3 and 4.

We use whole numbers to count how many things we have. This is why these numbers are also sometimes called the counting numbers.

The counting numbers are always positive, and so we call them the **positive whole numbers**.

Sometimes we have **none** of a thing (if we have eaten all our sweets, we have none left!) So, 0 (zero) is one of the counting numbers. Zero is a very special number. As we will see, in some ways it is different from other numbers.

## Numerals, digits and numbers

Numerals    A numeral is any single, positive figure.
A numeral can be on its own, or as part of a group of figures.

For example, 6 is a numeral on its own.
6 is one of four numerals that make up the group 3695.

Digit    'Digit' means the same as 'numeral'. We use both words in the same way.

For example, 6 is a digit on its own.
6 is one of four digits that make up the group 3695.

Number    We use the word 'number' to describe the value of any single digit (or numeral), or of any **combination** of digits (or numerals).

For example, 6 on its own can be called a number, a numeral, or a digit.
3695 is a number. It is made up of four digits (or numerals).

## What are the counting numbers?

As we learned earlier, the counting numbers are the positive whole numbers, and zero. Every number in the English system is made out of ten digits.

| 0 | | 1 | 2 | 3 | 4 | 5 | 6 | 7 | 8 | 9 |
|---|---|---|---|---|---|---|---|---|---|---|
| zero | | one | two | three | four | five | six | seven | eight | nine |
| (or nought) | | | | | | | | | | |

The first ten positive whole numbers are: 0, 1, 2, 3, 4, 5, 6, 7, 8, 9
The next ten positive whole numbers are: 10, 11, 12, 13, 14, 15, 16, 17, 18, 19
The next ten positive whole numbers are: 20, 21, 22, 23, 24, 25, 26, 27, 28, 29
… and so on!

NOTE: We will learn about the negative whole numbers later in the unit.

## Place value

Each of the ten digits has a value when it is on its own.

When we combine digits to make a number, the **value** of each digit in the number changes when this digit is in different places in the number. We call this the place value of a digit. For example:

In the number 3978, the digit 8 has a place value of 8.
In the number 5384, the digit 8 has a place value of 8 × 10 = 80.
In the number 4853, the digit 8 has a place value of 8 × 100 = 800.

## Reading and writing numbers in English

The number 543 is made up like this:

$$
\begin{array}{ll}
5 \times 100 = 500 & \text{(five hundred)} \\
+4 \times 10 = \phantom{0}40 & \text{(forty)} \\
+3 \times 1 = \phantom{00}3 & \text{(three)} \\
\hline
\phantom{0000}543 &
\end{array}
$$

We use this to tell us how to read a number.

The number 543 is read as 'five hundred and forty-three'.

If the number is bigger than 999, then we must split the digits in the number into groups of **three digits** – starting from the **right-hand side** of the number.

We then read each group separately. We always start reading numbers from the left-hand side. Each group of three digits is read in the same way.

For example, 763 is read as 'seven hundred and sixty-three'.

NOTE: After the name of any 'hundreds' and before the names of any 'tens' and 'ones', we add the word 'and' (but not in America!). We do this even if there are **no** 'hundreds'.

Each group of three digits (except the right-hand group) has an extra name:

| Group | 4 | 3 | 2 | 1 |
| --- | --- | --- | --- | --- |
| Extra name | billion | million | thousand | (no extra name) |

**Examples**

Write the digits in each number in groups of three to help you to write the number in words.

a) 2894

2 894, two **thousand**, eight hundred and ninety-four

b) 35766

35 766, thirty-five **thousand**, seven hundred and sixty-six

c) 834123

834 123, eight hundred and thirty-four **thousand**, one hundred and twenty-three

d) 5693402

5 693 402, five **million**, six hundred and ninety-three **thousand**, four hundred and two

e) 70250009

70 250 009, seventy **million**, two hundred and fifty **thousand**, and nine

f) 167304915

167 304 915, one hundred and sixty-seven **million**, three hundred and four **thousand**, nine hundred and fifteen

g) 2300056900

2 300 056 900, two **billion**, three hundred **million**, fifty-six **thousand**, nine hundred

**Exercise 1**

1 Write each of these numbers using figures.
   a) three hundred and ninety-six
   b) five thousand, and ten
   c) seventy thousand, two hundred
   d) nine million, two thousand, and fifty-one
   e) seven hundred and sixty-two million, five hundred and four thousand, and nineteen
   f) twenty million, two hundred and two thousand, and twenty
   g) twenty-five thousand, four hundred and sixty-three
   h) three hundred and seventy thousand
   i) one million, two hundred and four
   j) one thousand, two hundred and forty
   k) six hundred and thirty-three
   l) seventeen thousand, nine hundred and eighteen

**2** Write each of these numbers in words.
  a) 84
  b) 23590
  c) 93145670
  d) 764809
  e) 6049
  f) 9080004

**3** In the number 3<u>8</u>4, the place value of the underlined digit is 80.
Write down the place value of the underlined digit(s) in each of these numbers.
  a) 62<u>3</u>4
  b) 1<u>2</u>3 456 789
  c) 95<u>6</u>70
  d) <u>2</u>003
  e) 9<u>4</u> 705
  f) 423<u>6</u>
  g) <u>6</u>570
  h) 108<u>2</u>2
  i) <u>56</u>6 002
  j) 1 0<u>42</u> 240
  k) 8 <u>765</u> 432
  l) 21 5<u>36</u> 300 000

## Odd and even numbers

Half of the whole numbers can be divided exactly by 2 and give no **remainder**.

These numbers are called the **even** numbers: 2, 4, 6, 8, 10, 12, 14, 16, and so on.

The other half of the whole numbers cannot be divided exactly by 2.

These numbers are called the **odd** numbers: 1, 3, 5, 7, 9, 11, 13, 15, 17, and so on.

The number zero (0), is a special number. It is really **neither odd nor even**. However, it can be divided exactly by 2, (0 ÷ 2 = 0) and so it is sometimes included as an even number.

## Writing positive whole numbers in order

We know that all the whole numbers can be arranged in the following **order**: 0, 1, 2, 3, 4, 5, 6, ...

If we know one number, we can find the next whole number in the list by adding 1.

The **smallest** positive whole number is zero (0).

There is no **biggest** positive whole number (because we can always add 1 more!).

In maths, we use the symbol '>' to stand for 'is bigger than' or 'is more than'.

We use the symbol '<' to stand for 'is smaller than' or 'is less than'.

For example:

7 > 4 means '7 is bigger than 4' or '7 is more than 4'.
3 < 8 means '3 is smaller than 8' or '3 is less than 8'.

If we add an equals sign to these two symbols, we get two new symbols:

≥ means 'more than or equal to'.
≤ means 'less than or equal to'.

For example:

The positive whole numbers ≥ 5 are 5, 6, 7, 8, … and so on.
The positive whole numbers ≤ 5 are 5, 4, 3, 2, 1 and 0.
5 is part of both lists because of the 'or equal to'.

If we write a list of numbers in order, we can start with either the smallest number in the list or the biggest number in the list.

If we start with the smallest number, the numbers in the list will get bigger and bigger.

We say that the numbers are in ascending order.

Smallest number    ascending order    Biggest number

If we start with the biggest number, the numbers in the list will get smaller and smaller.

We say that the numbers are in descending order.

Biggest number    descending order    Smallest number

## Exercise 2

1 Look at these numbers: 97, 32, 23, 28, 302, 203
   a) Which is the biggest odd number?
   b) Which is the smallest even number?

2 Write down all the even numbers between 25 and 35.

3 Write down all the odd numbers that are > 84 but ≤ 99.

4  Write each group of numbers in ascending order.
   a) 74, 168, 39, 421
   b) 3842, 5814, 3874, 3801, 4765

5  Write each group of numbers in descending order.
   a) 399, 425, 103, 84, 429          b) 9434, 9646, 9951, 9653

6  Using the digits 4, 6, 7, 1 and 5 (do not use the same digit more than once)
   a) make the biggest five-digit number
   b) make the smallest five-digit number.

7  Using the digits 5, 4, 9 and 6 (do not use the same digit more than once)
   a) make the biggest four-digit even number
   b) make the smallest four-digit odd number.

8  a) Using the digits 8, 5, 4 and 3, make as many four-digit numbers as you
      can (use each digit only once in each number).
   b) Write your numbers in ascending order.
   c) How many numbers begin with the digit 8?

## The number line

We can think of whole numbers as points on a line. We call this line
a **number line**.

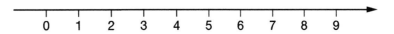

- Draw a line. Choose a point near the left end of the line and
  label it with the smallest number you want to show on the
  number line.
- Make marks at **equal** distances along the line from your first
  number.
- Label these marks (or points) with the other numbers in **ascending**
  order.

The arrow at the right-hand side of the line shows that there can be
more numbers along the line.

Numbers on a number line always get **bigger** as we move to the
**right**, and always get **smaller** as we move to the **left**. So, any number
on a number line is always bigger than all the numbers to its left, and
it is always smaller than all the numbers to its right.

Looking at the line above, we can say that $2 > 1$ and $6 < 7$.

## Addition of positive whole numbers

**Addition** is another word for adding. Here are two different methods we can use to **add** positive whole numbers.

### Writing numbers in columns

First write the numbers in neat **columns**, so that each digit with the same place value is in the same column.

Then add together the numbers in each of the columns, starting with the **units** (ones). If any of these answers comes to 10 or more, write down the **last** digit and 'carry' the other digit over to the next column. Add this digit together with the others in that column.

**Example**

Work out $4567 + 835$.

| | Thousands | Hundreds | Tens | Units |
|---|---|---|---|---|
| | 4 | 5 | 6 | 7 |
| + | | 8 | 3 | 5 |
| | 5 | 4 | 0 | 2 |
| | 1 | 1 | 1 | |

Units: $7 + 5 = 12$. Write 2 and carry the 1 to the next column.
Tens: $6 + 3 + 1$ (carried) $= 10$. Write 0 and carry the 1.
Hundreds: $5 + 8 + 1$ (carried) $= 14$. Write 4 and carry the 1.
Thousands: $4 + 1$ (carried) $= 5$.

### Using a number line

A number line shows a kind of 'picture' of the numbers and can help us to 'see' what happens when we add numbers together.

Add each digit separately, depending on its place value. Start with the biggest.

If you practise, you won't need to draw the number line every time!

**Example**

Work out $37 + 26$.

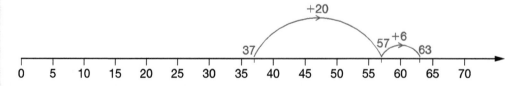

Add 20 first: $37 + 20 = 57$
(The digit '2' in the number 26 has a **place value** of 20.)
Then add 6: $57 + 6 = 63$
So $37 + 26 = 63$

## Word problems

When we solve word problems, we must first understand what the words are saying, and then think about how we can work out what the question asks.

There is no formula we can use to work out all word problems – each problem is a little different. However, we can always use our maths skills, together with some careful thought and common sense, to work out the answer.

Remember that **how** you work out the answer is just as important as the answer itself.

It is important to think clearly and work logically. Use these steps to help you.
- Make sure you understand what all the **words** mean. If you don't, look them up in a dictionary.
- Make sure you understand the whole **story** of the word problem. You could make a list of the key facts, draw a picture – or anything that helps you.
- Make sure you understand the **question** and know what you must work out.
- Decide how to work out the answer to the question. What **calculations** must you do? Use all the skills you have learned in maths and other subjects.
- When you have the maths answer, make sure that you answer the question using **words**.

**Example**

Alan is reading a book. On Monday he reads 153 pages. On Tuesday he reads 78 pages. On Wednesday he reads 136 pages and finishes the book.

How many pages does this book have?

We know how many pages Alan reads each day.
We must work out the total number of pages in the book – this means we must add the number of pages he reads each day.
So we must calculate: $153 + 78 + 136 = 367$
The book has 367 pages.

**Exercise 3**

**1** Draw a number line for each of these calculations and use it to work out the answer.
  a) 14 + 15　　　　　　　　　b) 13 + 18
  c) 7 + 36　　　　　　　　　　d) 18 + 25
  e) 24 + 29　　　　　　　　　f) 19 + 27

**2** What must you add to each of these numbers to make the answer equal to 100? You can use a number line to help you.
  a) 9 + _____ = 100　　　　　b) 96 + _____ = 100
  c) 45 + _____ = 100　　　　d) 37 + _____ = 100
  e) 62 + _____ = 100　　　　f) 83 + _____ = 100

**3** Add each of these sets of numbers by writing them in columns.
  a) 765 + 23　　　　　　　　　b) 27 + 56
  c) 76 + 98　　　　　　　　　　d) 324 + 628
  e) 1273 + 729　　　　　　　　f) 3495 + 8708
  g) 67 + 89 + 45　　　　　　　h) 934 + 715 + 86
  i) 431 + 865 + 245　　　　　j) 123 456 + 876 544
  k) 187 + 54 + 3210　　　　　l) 4801 + 2191 + 8463

**4** Work out the answer to each of these additions in your head.
  a) 15 + 12　　　　　　　　　　b) 24 + 32
  c) 19 + 16　　　　　　　　　　d) 37 + 17
  e) 26 + 48　　　　　　　　　　f) 29 + 41
  g) 63 + 57　　　　　　　　　　h) 13 + 99

**5** Work out the missing digits to make these additions correct.

a)
```
    2  4  9
 + □  □  □
 ──────────
    8  □  5
```

b)
```
       □  6  3
       7  □  2
 +     5  8  □
 ──────────────
    □  0  4  2
```

c)
```
       1  7  □  6
       □  1  5  4
 +     7  □  8  □
 ──────────────────
    □  1  3  3  8
```

d)
```
          9  7  6
    1  □  □  □
 +        6  7
 ──────────────
    □  4  8  6
```

**6** This signpost is between the cities of Poole and London.

How far is it from Poole to London?

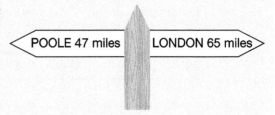

**7** What is the total cost of the computer package advertised here?

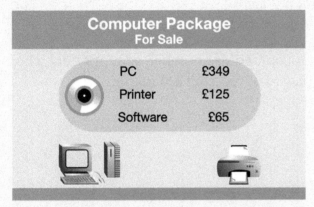

**8** What is the total cost of the holiday?

**9** The number of pupils in each class at Pinetown School is shown in the table below.

| Year 7 | | Year 8 | | Year 9 | |
|---|---|---|---|---|---|
| 7.1 | 25 | 8.1 | 27 | 9.1 | 26 |
| 7.2 | 29 | 8.2 | 30 | 9.1 | 25 |
| 7.3 | 29 | 8.3 | 25 | 9.3 | 27 |
| 7.4 | 30 | 8.4 | 24 | 9.4 | 29 |
| 7.5 | 28 | 8.5 | 26 | 9.5 | 26 |

a) How many pupils are in Year 7?
b) How many pupils are in Year 8?
c) How many pupils are in Year 9?
d) How many pupils are in the whole school?

10 The cinema at Big C has two screens. The table shows the number of tickets sold each day last week for Screen 1.

| Mon | Tues | Wed | Thurs | Fri | Sat | Sun |
|-----|------|-----|-------|-----|-----|-----|
| 125 | 87 | 95 | 105 | 278 | 487 | 201 |

a) How many tickets were sold for Screen 1 last week?
b) 891 tickets were sold for Screen 2 last week.
   What is the total number of tickets sold by the cinema last week?

11 Manchester United Football Club counted the number of supporters who watched the last four matches played at the club's stadium.

| Match | Number of supporters |
|-------|----------------------|
| 1 | 21 004 |
| 2 | 19 750 |
| 3 | 18 009 |
| 4 | 22 267 |

What is the total number of supporters that watched the last four matches played?

 ## Subtraction of positive whole numbers

When we **subtract** (take away) one number from another number, we can use 'columns' or a 'number line' in almost the same way that we used them to add numbers.

### Writing numbers in columns

Write the numbers in neat columns, so that each digit with the same place value is in the same column.

The **order** in which we write the numbers in columns for **subtraction** is very important.

Always write the number to be subtracted **below** the other number.

Start subtracting the numbers in the units (ones) column first. Continue with the tens, hundreds, and so on. Sometimes the number we want to subtract is bigger than the number we want to subtract it from. In this case, we 'borrow' 1 from the column on the left. When we move this 1 a column to the right, it will become 10 times bigger. This leaves one less in the column we have borrowed from.

**Example**

Work out $7238 - 642$.

| Thousands | Hundreds | Tens | Units |
|---|---|---|---|
| 6 | 11 | 1 | |
| 7 | 2̸ | 3 | 8 |
| − | 6 | 4 | 2 |
| 6 | 5 | 9 | 6 |

Units:      $8 - 2 = 6$

Tens:       $3 - 4$ cannot be done. Borrow 1 from the 2 in the hundreds column (leaving 1). 1 from the hundreds column is the same as 10 tens, so now our calculation is: $(3 + 10) - 4 = 9$

Hundreds:   $1 - 6$ cannot be done. Borrow 1 from the 7 is the thousands column (leaving 6). 1 from the thousands column is the same as 10 hundreds, so now our calculation is: $(1 + 10) - 6 = 5$

Thousands:  $6 - 0 = 6$

NOTE: Addition is the 'opposite' operation to subtraction, so we can use addition to check that a subtraction calculation is correct: $6596 + 642 = 7238$ ✓

Sometimes we need to borrow but the number in the column we want to borrow from is 0.

For example, we might want to borrow from the tens column, but the tens digit is 0. When this happens, we borrow 1 from the next column along (the hundreds column). This gives us '10' in the tens column. We can then borrow 1 from this 10, leaving 9.

**Examples**

Work out these.

a) $1762 - 873$

| | 0 | 16 | 15 | 1 |
|---|---|---|---|---|
| | 1̸ | 7 | 6̸ | 2 |
| − | | 8 | 7 | 3 |
| | | 8 | 8 | 9 |

b) $3006 - 1847$

| | 2 | 9 | 9 | 1 |
|---|---|---|---|---|
| | 3̸ | 0̸ | 0̸ | 6 |
| − | 1 | 8 | 4 | 7 |
| | 1 | 1 | 5 | 9 |

c) $9012 - 5678$

| | 8 | 9 | 10 | 1 |
|---|---|---|---|---|
| | 9̸ | 0̸ | 1̸ | 2 |
| − | 5 | 6 | 7 | 8 |
| | 3 | 3 | 3 | 4 |

**Using a number line**

We can use a number line for subtraction in exactly the same way as we used one for addition. The only difference is that the movement on the line will be to the **left** (smaller numbers) when we are **subtracting**.

## Exercise 4

**1** Draw a number line for each of these calculations and use it to work out the answer.

  a) $56 - 28$  b) $64 - 18$

  c) $44 - 19$  d) $73 - 38$

**2** Work out these subtractions in your head.

  a) $100 - 95$  b) $100 - 8$

  c) $100 - 57$  d) $100 - 32$

  e) $100 - 24$  f) $100 - 83$

  g) $100 - 41$  h) $100 - 79$

**3** Work out these subtractions in your head.

  a) $26 - 10$  b) $26 - 9$

  c) $87 - 37$  d) $87 - 38$

  e) $200 - 110$  f) $204 - 99$

  g) $500 - 350$  h) $500 - 199$

  i) $1000 - 425$  j) $1003 - 999$

**4** Work out these subtractions by writing them in columns.
  Use addition to check your answers.

  a) $978 - 624$  b) $843 - 415$

  c) $1754 - 470$  d) $407 - 249$

  e) $5070 - 2846$  f) $2345 - 1876$

  g) $8045 - 1777$  h) $10\,000 - 6723$

  i) $60\,152 - 1895$  j) $91\,346 - 88\,978$

**5** Work out the missing digits to make these subtractions correct.

  a)
$$
\begin{array}{r}
2\ \square\ 4 \\
-\ \square\ 9\ \square \\
\hline
8\ \ \ 8
\end{array}
$$

  b)
$$
\begin{array}{r}
7\ \square\ 2 \\
-\ \square\ 3\ 9 \\
\hline
4\ \ 8\ \square
\end{array}
$$

  c)
$$
\begin{array}{r}
\square\ \square\ 9\ 4 \\
-\ \ \ 9\ \square\ 6 \\
\hline
6\ \ 8\ \square
\end{array}
$$

  d)
$$
\begin{array}{r}
\square\ 5\ 3\ \square\ 1 \\
-\ \square\ 4\ 0\ \square \\
\hline
4\ 9\ \square\ 1\ 8
\end{array}
$$

**6** A car park has spaces for 345 cars. One Tuesday, 256 spaces are used. How many spaces are still empty?

**7** A school has 843 pupils. There are 459 girls. How many boys are in this school?

**8** Every car has a meter to measure how many kilometres it travels.
This meter was read in three different cars (A, B and C) at the beginning of the year, and again at the end of the year. The table shows the results.

| | Beginning of the year | End of the year |
|---|---|---|
| Car A | 2 501 | 10 980 |
| Car B | 55 667 | 67 310 |
| Car C | 48 050 | 61 909 |

Which car has travelled the most number of kilometres during this year?

**9** The table shows the prices of some electrical goods advertised at a shop.

| Item | Cost |
|---|---|
| TV | $199 |
| Video | $250 |
| Combined TV/video | $375 |
| Music system | $185 |

a) If I buy the TV and video separately, I will pay **more** than if I buy the combined TV/video. How much more will I pay?

b) I want to buy the combined TV/video and the music system. I have saved $500. How much more money do I still need to save before I can buy them?

 ## Multiplication of positive whole numbers

**Multiplication** is really like adding the same number together again and again as many times as we need. For example, we could say that 3 × 4 is the same as adding together four 3s (3 + 3 + 3 + 3 = 12), or adding together three 4s (4 + 4 + 4 = 12).

When the numbers get bigger, this method takes a lot of time, and so it is very important that you know the multiplication **tables** in your head. You should learn up to at least 12 × 12 (see page 16).

| ×  | 1  | 2  | 3  | 4  | 5  | 6  | 7  | 8  | 9   | 10  | 11  | 12  |
|----|----|----|----|----|----|----|----|----|-----|-----|-----|-----|
| 1  | 1  | 2  | 3  | 4  | 5  | 6  | 7  | 8  | 9   | 10  | 11  | 12  |
| 2  | 2  | 4  | 6  | 8  | 10 | 12 | 14 | 16 | 18  | 20  | 22  | 24  |
| 3  | 3  | 6  | 9  | 12 | 15 | 18 | 21 | 24 | 27  | 30  | 33  | 36  |
| 4  | 4  | 8  | 12 | 16 | 20 | 24 | 28 | 32 | 36  | 40  | 44  | 48  |
| 5  | 5  | 10 | 15 | 20 | 25 | 30 | 35 | 40 | 45  | 50  | 55  | 60  |
| 6  | 6  | 12 | 18 | 24 | 30 | 36 | 42 | 48 | 54  | 60  | 66  | 72  |
| 7  | 7  | 14 | 21 | 28 | 35 | 42 | 49 | 56 | 63  | 70  | 77  | 84  |
| 8  | 8  | 16 | 24 | 32 | 40 | 48 | 56 | 64 | 72  | 80  | 88  | 96  |
| 9  | 9  | 18 | 27 | 36 | 45 | 54 | 63 | 72 | 81  | 90  | 99  | 108 |
| 10 | 10 | 20 | 30 | 40 | 50 | 60 | 70 | 80 | 90  | 100 | 110 | 120 |
| 11 | 11 | 22 | 33 | 44 | 55 | 66 | 77 | 88 | 99  | 110 | 121 | 132 |
| 12 | 12 | 24 | 36 | 48 | 60 | 72 | 84 | 96 | 108 | 120 | 132 | 144 |

The better you know these tables, the quicker and easier you will find all of mathematics!

## Short multiplication

**Short multiplication** is any calculation when the **multiplying number** is less than 10, such as $165 \times 7$.

To work out a short multiplication, multiply each digit in turn. Start with the units (ones), then continue with the tens, the hundreds and so on. Remember the place value of each digit when you are working.

**Example**

Calculate $165 \times 7$.

| Thousands | Hundreds | Tens | Units |
|-----------|----------|------|-------|
|           | 1        | 6    | 5     |
| ×         |          |      | 7     |
| 1         | 1        | 5    | 5     |
| 1         | 4        | 3    |       |

| Units:     | $7 \times 5 = 35$. Write 5 and carry 3 to the next column.                                         |
|------------|----------------------------------------------------------------------------------------------------|
| Tens:      | $7 \times 6 = 42$ and $42 + 3$ (carried) $= 45$. Write 5 and carry 4.                               |
| Hundreds:  | $7 \times 1 = 7$ and $7 + 4 = 11$. Write 1 and carry 1.                                             |
| Thousands: | There are now no more digits to multiply, so the last carried digit becomes 1 thousand.            |

**Examples**

Calculate these.

a)  162 × 4

```
      1   6   2
  ×           4
  ─────────────
      6   4   8
      2
```

b)  9071 × 7

```
          9   0   7   1
  ×                   7
  ─────────────────────
      6   3   4   9   7
                  4
```

c)  4835 × 8

```
      4   8   3   5
  ×               8
  ─────────────────
      3   8   6   8   0
          6   2   4
```

## Multiplying a whole number by 10, 100, 1000, …

- When we multiply a number by **10**, the answer is the same as writing an extra **0** on the end of the number we started with.
- When we multiply a number by **100**, the answer is the same as writing an extra **two 0s** on the end of the number we started with. (Remember: 100 = 10 × 10)
- When we multiply a number by **1000**, the answer is the same as writing an extra **three 0s** on the end of the number we started with. (Remember: 1000 = 10 × 10 × 10)
- And so on … Can you see the pattern?

In general, to multiply a number by 10, 100, 1000 and so on, count the number of 0s in the multiplying number, and write this number of 0s on the end of the other number. For example:

782 × 10 = 7820
963 × 100 = 96 300

9274 × 1000 = 9 274 000
15 × 1 000 000 = 15 000 000

## Multiplying a whole number by other numbers ending in a 0

Now we know how to multiply by 10, 100, 1000 and so on, we can use this to help us multiply by numbers such as 20, 30, 40 and so on.

First, write the multiplying number as 10 × some other number. For example, 20 = 10 × 2.

Now, do the two multiplications separately.

**Example**

Work out $674 \times 20$.

$$\begin{aligned} 674 \times 20 &= 674 \times 10 \times 2 \quad \text{First multiply by 10.} \\ &= 6740 \times 2 \quad\quad \text{Then multiply by 2.} \\ &= 13\,480 \end{aligned}$$

**Exercise 5**

1 Work out the answer to each of these short multiplications.
   a) $21 \times 4$                      b) $17 \times 5$
   c) $36 \times 7$                      d) $183 \times 3$
   e) $264 \times 8$                     f) $3179 \times 5$
   g) $4012 \times 6$                    h) $6012 \times 7$

2 Write down the answer to each of these multiplications.
   a) $132 \times 10$                    b) $123 \times 100$
   c) $47 \times 1000$                   d) $384 \times 100$

3 Work out the missing numbers to make these multiplications correct.
   a) $231 \times 10 = \square$          b) $\square \times 1000 = 514\,000$
   c) $172 \times \square = 17\,200$

4 I can buy eight doughnuts for $1. How many doughnuts can I buy for $5?

5 A packet contains 24 biscuits. This packet costs 39 cents ($0.39).
   I buy 10 packets.
   a) How many biscuits do I buy?
   b) How much do the 10 packets cost me?

6 In each maths classroom there are 30 chairs, 20 tables, 50 pencils and 40 rulers.
   There are seven rooms in the maths department. For the whole maths
   department, find the total number of
   a) chairs                            b) tables
   c) pencils                           d) rulers.

7 60 seconds = 1 minute        60 minutes = 1 hour
   Calculate the number of seconds in
   a) 5 minutes                         b) 7 minutes
   c) 20 minutes                        d) 2 hours.

8 There are 32 classes in a school. Each class has 30 pupils. How many pupils
   are there in this school?

9 The distance all round the athletics running track is 400 metres. How far
   would you run if you ran around the track 25 times?

**10** Use the quickest way you can to calculate these multiplications.

a) $253 \times 30$                    b) $357 \times 20$

c) $537 \times 40$                    d) $615 \times 20$

e) $186 \times 70$                    f) $239 \times 90$

g) $412 \times 80$                    h) $142 \times 70$

i) $632 \times 30$                    j) $260 \times 50$

k) $253 \times 60$                    l) $6795 \times 80$

##  Division of positive whole numbers

Division is the 'opposite' process of multiplication. When we **divide** a number, we separate it into smaller, equal parts (or numbers).

Learning the multiplication table on page 16 will also help you with division.

### Short division

**Short division** is any calculation when the number we are **dividing by** is less than 10, such as $882 \div 7$.

To work out a short division, divide each digit in turn. Start at the left of the number and work towards the right.

**Example**      Work out $882 \div 7$.

$$\begin{array}{r} 1\,2\,6 \\ 7\overline{)8^1 8^4 2} \end{array}$$

Hundreds:  $8 \div 7 = 1$ remainder 1
Write 1 above the 8 and carry the remainder 1 to the next column.

Tens:  $18 \div 7 = 2$ remainder 4
Write 2 above the 8 and carry the 2 to the next column.

Units:  $42 \div 7 = 6$ exactly.
There is no remainder this time.

So $882 \div 7 = 126$

NOTE: Multiplication is the 'opposite' operation to division, so we can use multiplication to check that a division calculation is correct:
$126 \times 7 = 882$ ✓

**Examples**      a)  $$\begin{array}{r} 2\,4\,5 \\ 6\overline{)14^2 7^3 0} \end{array}$$          b)  $$\begin{array}{r} 3\,0\,7 \\ 9\overline{)276^6 3} \end{array}$$

## Dividing a whole number by 10, 100, 1000, ...

Division is the 'opposite' operation to multiplication, so dividing a whole number by 10, 100, 1000 and so on is the 'opposite' to multiplying by 10, 100, 1000 and so on.

In general, to divide a number by 10, 100, 1000 and so on, count the number of 0s in the dividing number, and remove (cross out) this number of 0s from the end of the other number. For example:

$8640 \div 10 = 864$
$23\,100 \div 100 = 231$
$429\,000 \div 1000 = 429$

## Dividing a whole number by other numbers ending in a 0

Now we know how to divide by 10, 100, 1000 and so on, we can use this to help us divide by numbers such as 20, 30, 40 and so on.

For example, dividing by 30 is the same as dividing by 10 and then dividing by 3; dividing by 50 is the same as dividing by 10 and then dividing by 5.

**Example**

Work out $7530 \div 30$.

$$7530 \div 30 = 7530 \div 10 \div 3 \qquad \text{First divide by 10.}$$
$$= 753 \div 3 \qquad\qquad \text{Then divide by 3.}$$
$$= 251$$

### Exercise 6

1 Calculate these divisions. Show your working clearly. Give the remainder if there is one. (Use multiplication to check your answers.)
   a) $85 \div 5$          b) $471 \div 3$          c) $816 \div 6$
   d) $455 \div 6$          e) $3146 \div 8$          f) $824 \div 4$
   g) $9882 \div 9$          h) $80\,560 \div 4$

2 Write down the answer to each of these divisions.
   a) $45\,600 \div 10$          b) $465\,000 \div 1000$          c) $64\,000 \div 1000$
   d) $65\,400 \div 100$          e) $123\,000 \div 10$          f) $17\,000 \div 100$

**3** Work out the missing numbers to make these divisions correct.
  a) $56\,400 \div \square = 564$      b) $\square \div 1000 = 702$
  c) $35\,000 \div \square = 3500$

**4** Use the quickest way you can to calculate these divisions.
  a) $7590 \div 30$      b) $7110 \div 90$      c) $21\,480 \div 40$
  d) $7560 \div 60$      e) $900 \div 20$      f) $30\,650 \div 50$
  g) $13\,000 \div 500$    h) $263\,700 \div 900$    i) $329\,600 \div 800$

**5** The picture shows the prices of some maths equipment.

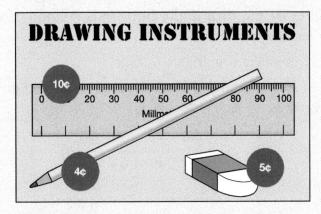

I have 80 cents to spend. I spend all the money on only one kind of thing.
  a) How many pencils can I buy?
  b) How many erasers can I buy?
  c) How many rulers can I buy?

**6** A minibus holds 20 passengers. How many minibuses will I need to carry 160 passengers?

**7** How many €50 notes will I need to pay a bill of €750?

**8** A bus can hold 70 passengers. How many buses will I need to take 840 pupils on a trip?

**9** A class studies maths for 300 minutes each week. If each lesson is 60 minutes long, how many lessons do they have in a week?

**10** How many minutes are there in 420 seconds?

## Long multiplication

When the multiplying number is **bigger than 10**, we use the method of **long multiplication**.

### The long multiplication process

1 Multiply the number by the digit in the **units** column of the multiplying number, then

2 Multiply the number by the digit in the **tens** column of the multiplying number, then

3 Multiply the number by the digit in the **hundreds** column of the multiplying number, and so on …

4 Then **add** all these answers together.

NOTE: It is important to remember the **place value** of each digit in the multiplying number.

**Examples**

Work out these.

a) $24 \times 17$

$$
\begin{array}{r}
2\ 4 \\
\times\quad 1\ 7 \\
\hline
1\ 6\ 8 \quad \leftarrow 24 \times 7 \\
2\ 4\ 0 \quad \leftarrow 24 \times 10 \\
\hline
4\ 0\ 8
\end{array}
$$

b) $145 \times 62$

$$
\begin{array}{r}
1\ 4\ 5 \\
\times\quad 6\ 2 \\
\hline
2\ 9\ 0 \quad \leftarrow 145 \times 2 \\
8\ 7\ 0\ 0 \quad \leftarrow 145 \times 60 \\
\hline
8\ 9\ 9\ 0
\end{array}
$$

c) $273 \times 234$

$$
\begin{array}{r}
2\ 7\ 3 \\
\times\quad 2\ 3\ 4 \\
\hline
1\ 0\ 9\ 2 \quad \leftarrow 273 \times 4 \\
8\ 1\ 9\ 0 \quad \leftarrow 273 \times 30 \\
5\ 4\ 6\ 0\ 0 \quad \leftarrow 273 \times 200 \\
\hline
6\ 3\ 8\ 8\ 2
\end{array}
$$

## Long division

**Long division** works in exactly the same way as short division, except that we write down all the working out steps. Remember to start at the **left** of the number and work towards the **right**.

### The long division process

1 Look at the first digit in the number you want to divide. How many times will the dividing number divide into it? (You can ignore any remainder for now.) Write your answer above the division line.

NOTE: If the dividing number is bigger than the first digit, you will need to look at the first two (or more) digits together.

2 Multiply the number you are dividing by the number you have just written above the division line. Write the answer below the digit(s) you started with.

3 Subtract the answer to your multiplication from the digit(s) you started with. This gives the remainder.

4 Bring down the next digit from the number you are dividing, and write it next to the remainder.

5 Repeat steps 1 to 4 until you have no more digits to bring down from the number you are dividing.

**Example**

Work out $952 \div 7$ using long division.

```
      1 3 6
  7 ) 9 5 2
    − 7
      2 5
    − 2 1
      4 2
    − 4 2
        0
```

Hundreds: 1 How many times will 7 divide into 9?
$9 \div 7 = 1$ plus a remainder.
Write 1 above the 9. This is the first digit of the answer.
2 $1 \times 7 = 7$. Write 7 below the 9.
3 $9 - 7 = 2$. This is the remainder.
4 Bring down the next digit (5) from the number we are dividing. This makes 25.

Tens: 1 How many times will 7 divide into 25?
$25 \div 7 = 3$ plus a remainder.
Write 3 above the 5. This is the second digit of the answer.
2 $3 \times 7 = 21$. Write 21 below the 25.
3 $25 - 21 = 4$. This is the remainder.
4 Bring down the next digit (2) from the number we are dividing. This makes 42.

Units: 1 How many times will 7 divide into 42?
$42 \div 7 = 6$ exactly.
Write 6 above the 2. This is the final digit of the answer.
2 $6 \times 7 = 42$. Write 42 below the 42.
3 $42 - 42 = 0$. There is no remainder.
4 There are no more digits to bring down.
So $952 \div 7 = 136$

In the above example, we divided by a single digit so that we could see how long division works. Usually we use long division only when the dividing number is **bigger than 10**.

Sometimes the answer to a long division will include a **remainder**.

Work out $45\,963 \div 17$.

```
      (0)2 7 0 3
  17) 4 5 9 6 3
  –   3 4
      ———
      1 1 9
  –   1 1 9
      ———
        0 6 3
  –      5 1
        ———
          1 2
```

**Ten thousands:**

1 17 will not divide into 4 so the first digit of the answer is 0. We do not normally write 0 at the beginning of a number.

**Thousands:**

1 How many times will 17 divide into 45? $2 \times 17 = 34$ (too small), $3 \times 17 = 51$ (too big). So $34 \div 17 = 2$ plus a remainder. Write 2 above the 5. This is the first digit of the answer.

2 $2 \times 17 = 34$. Write 34 below the 45.

3 $45 - 34 = 11$. This is the remainder.

4 Bring down the next digit (9) from the number we are dividing. This makes 119.

**Hundreds:**

1 How many times will 17 divide into 119? $7 \times 17 = 119$ exactly. Write 7 above the 9. This is the second digit of the answer.

2 $7 \times 17 = 119$. Write 119 below the 119.

3 $119 - 119 = 0$. There is no remainder, but we haven't finished yet so write down 0.

4 Bring down the next digit (6) from the number we are dividing.

**Tens:**

1 17 will not divide into 6 so the third digit of the answer of 0. Write 0 above the 6. This is important – if we don't, we will get the wrong answer! Bring down the next digit (3) from the number we are dividing.

**Units:**

1 How many times will 17 divide into 63? $3 \times 17 = 51$ (too small), $4 \times 17 = 68$ (too big). So $63 \div 17 = 3$ plus a remainder. Write 3 above the 3. This is the fourth digit of the answer.

2 $3 \times 17 = 51$. Write 51 below the 63.

3 $63 - 51 = 12$. This is the remainder.

4 There are no more digits to bring down.

So $45\,963 \div 17 = 2703$ remainder 12

NOTE: We will learn how to continue this division process to give a different kind of answer in Unit 4. For now, we will leave the answer as a whole number and a remainder.

## Exercise 7

**1** Use the long multiplication process to work out these multiplications.
a) 42 × 32
b) 76 × 32
c) 143 × 34
d) 265 × 42
e) 718 × 54
f) 1038 × 74
g) 765 × 451
h) 9852 × 672

**2** Use the long division process you have learned to work out these divisions.
a) 7871 ÷ 17
b) 4582 ÷ 29
c) 9471 ÷ 77
d) 4864 ÷ 19
e) 7560 ÷ 15
f) 20 928 ÷ 32
g) 11 368 ÷ 28
h) 11 232 ÷ 54
i) 34 977 ÷ 23
j) 3796 ÷ 33
k) 243 667 ÷ 142
l) 17 578 ÷ 117

**3** To travel on the bus from London to Oxford costs £29 for each person.
48 people are on the bus today. How much money was paid altogether?

**4** A train has 12 carriages. 118 passengers can sit in each carriage.
How many passengers can sit on the whole train?

**5** The picture shows the prices of some office furniture.

a) What is the cost of 14 desks?
b) What is the cost of 23 filing cabinets?
c) What is the total cost of 14 desks, 23 filing cabinets and 17 chairs?

**6** Work out these long divisions. Give the remainder in each case.
a) 410 ÷ 25
b) 607 ÷ 24
c) 800 ÷ 45
d) 525 ÷ 37

**7** Half a litre of milk costs 29 cents ($1 = 100 cents).
a) I have $5.00. How many litres of milk can I buy?
b) How much change (money back) will I get?

**8** Tins of tuna are packed in boxes of 24 tins. The local supermarket has 1000 tins of tuna.

a) How many full boxes is this?

b) How many tins will be left over?

## Properties of positive whole numbers

For some calculations, it doesn't matter in what order the numbers are. Check for yourselves that each of these properties is true by trying some more examples.

- We can add whole numbers in any order – the answer will be the same. For example:

  $2 + 3 = 5$ and $3 + 2 = 5$

  $1 + 4 + 5 = 10$, $1 + 5 + 4 = 10$, $4 + 1 + 5 = 10$, $4 + 5 + 1 = 10$, $5 + 1 + 4 = 10$ and $5 + 4 + 1 = 10$

Is this also true for subtraction? Check!

- We can multiply whole numbers in any order – the answer will be the same. For example:

  $2 \times 4 = 8$ and $4 \times 2 = 8$

  $2 \times 3 \times 4 = 24$, $2 \times 4 \times 3 = 24$, $3 \times 2 \times 4 = 24$, $3 \times 4 \times 2 = 24$, $4 \times 2 \times 3 = 24$ and $4 \times 3 \times 2 = 24$

Is this also true for division? Check!

- When we add numbers and then multiply the sum by another number, the answer is the same as multiplying each of the numbers first, and then adding the products. For example:

  $2 \times (3 + 4) = 2 \times 7 = 14$

  $2 \times (3 + 4) = (2 \times 3) + (2 \times 4) = 6 + 8 = 14$

Is this also true for multiplication and subtraction (instead of addition)? Check!

**Exercise 8**

**1** Copy and complete each **equation** (statement), writing + or × in each space to make it true.

a) $8 \square 7 = 7 \square 8 = 56$

b) $47 \square 14 = 14 \square 47 = 61$

c) $275 \square 24 = 24 \square 275 = 299$

d) $7 \square 16 = 16 \square 7 = 112$

e) $23 \square 9 = 9 \square 23 = 207$

f) $124 \square 297 = 297 \square 124 = 421$

**2** Write a number in each space to make the equation (statement) true.

a) $8 \times (7 + \square) = 8 \times 7 + 8 \times 9$

b) $11 \times (3 + 6) = 11 \times \square + 11 \times \square$

c) $10 \times 13 + 5 \times 13 = (10 + 5) \times \square$

d) $3 \times (\square - 7) = (3 \times 8) - (3 \times 7)$

e) $(9 \times 15) - (6 \times 15) = (9 - 6) \times \square$

f) $(4 - 3) \times 13 = (\square \times 13) - (3 \times \square)$

## The order of operations

Sometimes it is not very clear in which order to do a series of operations.

For example, does $16 - 2 + 9$ mean $(16 - 2) + 9 = 14 + 9 = 23$
or $16 - (2 + 9) = 16 - 11 = 5$?

does $4 + 3 \times 5$ mean $(4 + 3) \times 5 = 7 \times 5 = 35$
or $4 + (3 \times 5) = 4 + 15 = 19$?

In each case, the two different calculations give different answers, so we need to know which one is wanted.

This is why there are very clear rules that tell us in what order to do the calculations.

**1** If there are any **brackets** in the calculation, do the calculation inside the brackets first. You can then remove the brackets. Do not write any brackets if there are none already written! For example:

$4 + (9 - 6) = 4 + 3 = 7$
$7 \times (4 + 5) = 7 \times 9 = 63$

**2** If there are brackets **inside** other brackets, do the calculation in the brackets that are 'most inside' first. For example:

$2 + [14 - (2 + 7)] = 2 + [14 - 9] = 2 + 5 = 7$

**3** If there are **only additions and subtractions** in the calculation, work from left to right with each part in turn. For example:

$33 + 16 - 7 = 49 - 7 = 42$

**4** If there are **only multiplications and divisions** in the calculation, work from left to right. For example:

$144 \div 3 \times 12 = 48 \times 12 = 576$

**5** If the calculation contains any **combination** of addition and/or subtraction together with multiplication and/or division, then do all the multiplications and/or divisions **before** the additions and/or subtractions. For example:

$11 + 4 \times 7 - 26 = 11 + 28 - 26 = 39 - 26 = 13$

**Examples**   Work out these.

a) $4 + 3 \times 5$

$\quad 4 + 3 \times 5 = 4 + 15 = 19$

b) $10 \div 2 + 3$

$\quad 10 \div 2 + 3 = 5 + 3 = 8$

c) $10 \div (2 + 3)$

$\quad 10 \div (2 + 3) = 10 \div 5 = 2$

d) $(5 + 6) \times 3 + 4$

$\quad (5 + 6) \times 3 + 4 = 11 \times 3 + 4 = 33 + 4 = 37$

## Exercise 9

**1** Work out the answer to each of these, using the correct order of operations.

a) $7 + 6 \times 5$

b) $7 - (6 - 2)$

c) $24 \div 6 + 5$

d) $7 \times 6 + 8 \times 2$

e) $10 \div 5 + 8 \div 2$

f) $(5 - 2) \times 7 + 9$

g) $60 \div (5 + 7)$

h) $60 \div 5 + 7$

i) $4 \times 3 + 2$

j) $4 \times (3 + 2)$

k) $12 \times (20 - 2) \div 9$

l) $36 \div (5 + 4)$

m) $4 \times 12 \div 8 - 6$

n) $9 \times 9 - 5 \times 5$

o) $(9 + 5) \times (9 - 5)$

**2** Copy and complete each equation (statement) to make it true. Write $+$, $-$ or $\times$ in each space.

a) $5 \square 6 \square 7 = 37$

b) $5 \square 6 \square 7 = 47$

c) $15 \square 8 \square 9 = 87$

d) $15 \square 8 \square 9 = 129$

e) $15 \square 8 \square 9 = 111$

f) $15 \square 5 \square 3 = 6$

g) $5 \square 24 \square 6 = 1$

h) $19 \square 19 \square 7 \square 0 = 1$

i) $4 \square 4 \square 7 \square 2 = 30$

**3** Copy and complete each equation (statement) to make it true. Write $>$, $<$ or $=$ in each space

a) $8 \times 4 - 13 \square 19$

b) $4 \times 12 + 7 \times 5 \square 60$

c) $63 \div 9 \times 3 - 8 + 11 \square 25$

d) $64 \div (16 - 72 \div 6) \square 11$

e) $3 \times [3 + 2 \times (3 + 4)] \square 50$

f) $18 \times 6 \times 9 \div 12 \square 81$

g) $3 \times 21 \div 7 + 114 \div 6 \square 36$

h) $2448 \div (5 \times 3 - 4 + 6) \square 44$

i) $91 + 33 \times 7 - 136 \div 8 \square 315$

j) $(9 \times 7 - 117 \div 13) \times 3 - 63 \square 100$

**4** Write one pair of brackets in each of these calculations to make it correct.
For example, $13 - 7 - 4 = 10$ if we add these brackets:
$13 - (7 - 4) = 13 - 3 = 10$
a) $14 - 8 - 3 = 9$      b) $2 \times 7 + 8 = 30$
c) $2 \times 7 + 3 \times 3 = 32$      d) $2 \times 7 + 3 \times 3 = 60$
e) $2 \times 7 + 3 \times 3 = 51$      f) $4 \times 7 - 3 \times 5 = 80$

**5** Work out the answer to each of these calculations, using the correct order of operations.
a) $[(11 + 14) \times 4 - 5] \div 19$
b) $(30 + 40) \times [60 \div (40 - 30)]$
c) $[70 \times 2 + (60 - 50) \times 4] \div 9$
d) $8000 \div [1600 \div 20 \div 10 \times (40 + 10)]$
e) $[5 \times 52 - (5 \times 52 + 5 \times 36) \div 2] \div 5$
f) $[(453 + 33) \div 6 \times 9] - (63 - 49) \times 7$
g) $[345 - (144 - 121) \times 7] + (47 - 19) \times 5$
h) $75 - 38 \div 2 + 75 \div 5 \times 7 + 81 \div 3 \div 9 \times 7 - 15 + 6 \times 7$

**6** Use brackets and/or the signs $+$, $-$, $\times$ or $\div$ to make each of these calculations true. You may use as many brackets as you like. The signs can be in any order or combination. The first one has been done for you.
a) $6 \boxminus 3 \boxplus 2 \boxtimes 1 = 1$
b) $6 \square 3 \square 2 \square 1 = 2$
c) $6 \square 3 \square 2 \square 1 = 3$
d) $6 \square 3 \square 2 \square 1 = 4$
e) $6 \square 3 \square 2 \square 1 = 5$
f) $6 \square 3 \square 2 \square 1 = 6$
g) $6 \square 3 \square 2 \square 1 = 7$
h) $6 \square 3 \square 2 \square 1 = 8$
i) $6 \square 3 \square 2 \square 1 = 9$
j) $6 \square 3 \square 2 \square 1 = 10$

**7** The school caretaker has set out 17 rows of chairs. There are 15 chairs in each row. 280 people need to sit down. How many more chairs are needed?

**8** The admission charges to London Zoo are £4 for a child and £7 for an adult. Jane is taking some people to the zoo and has worked out that the total cost will be £336 for all of them. She has collected £84 from all the adults.
a) How many children are going to the zoo?
b) What is the total number of people going to the zoo?

## B Negative whole numbers and integers

For every **positive whole number** there is a 'matching' negative whole number. For example, the positive whole number +5 has a 'matching' negative whole number, −5. This is read as 'minus five' or as 'negative five'.

A negative whole number is made up of two parts:
- a negative sign
- one or more numerals, or digits.

So −5 is a number. That number is made up of the negative sign and the numeral, or digit, 5.

Together, the negative whole numbers, zero, and the positive whole numbers, are called integers.

We often write **positive** numbers without the **+ sign** (+5 is the same as 5), but we must always write the **− sign** with **negative** numbers (−5 is not the same as 5).

We have already used a **number line** to show zero and the positive whole numbers. We can extend this number line to show the negative whole numbers as well.

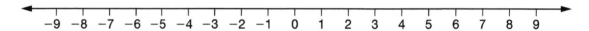

We need to use negative numbers in many real-life situations.
- In many countries the weather gets very cold. Temperatures below 0 °C are written as **negative temperatures**.
- If people spend more money than they have in their bank account, the balance of the account will be written as a **negative amount of money**.
- Sea level is always thought of as the 0 point of height on earth. A diver who goes down below the surface of the sea will record his position in **negative metres**.

Can you think of other situations when negative numbers are used?

### Writing integers in order

Remember that the numbers on a number line always get **bigger** towards the **right**, and **smaller** towards the **left**. This is still true when we show negative numbers on the number line as well.

So, −5 is **smaller** than −4 even though the numeral 5 is bigger than the numeral 4.

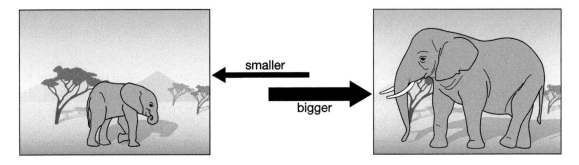

For example:

−5 is smaller than −1.
−3 is smaller than 1.
−4 is smaller than 0.
2 is smaller than 4.

2 is bigger than −3.
−4 is bigger than −5.
0 is bigger than −3.
−5 is bigger than −9.

With temperature, the **colder** temperatures (smaller numbers) are on the **left** of the number line (or thermometer). The **warmer** temperatures (bigger numbers) are on the **right**.

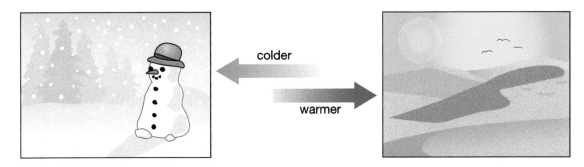

For example:

−5 °C is colder than −1 °C.
−3 °C is colder than 1 °C.
−4 °C is colder than 0 °C.
2 °C is colder than 4 °C.

2 °C is warmer than −3 °C.
−4 °C is warmer than −5 °C.
0 °C is warmer than −3 °C.
−5 °C is warmer than −9 °C.

**Examples**

a) List these temperatures from coldest to warmest. (This will be ascending order.)

3 °C, 5 °C, −2 °C, 0 °C, −4 °C

−4 °C, −2 °C, 0 °C, 3 °C, 5 °C

b) List these numbers in descending order.

50, −41, −18, −11, 28, 9

50, 28, 9, −11, −18, −41

**Exercise 10**

**1** Copy and complete each sentence. Write 'colder' or 'warmer' in each space to make the sentence correct.

a) −2 °C is _____ than −5 °C.
b) −1 °C is _____ than 4 °C.
c) 2 °C is _____ than −4 °C.
d) −10 °C is _____ than −5 °C.

**2** Copy and complete each sentence. Write 'more' or 'less' in each space to make the sentence correct.

a) −3 is _____ than 2.
b) 1 is _____ than −5.
c) −4 is _____ than −1.
d) −4 is _____ than −10.

**3** The table shows the temperatures in some cities around the world on Christmas Day.

| City | Temperature on Christmas Day |
| --- | --- |
| Edinburgh | −7 °C |
| London | 0 °C |
| Moscow | −22 °C |
| New York | −17 °C |
| Rome | 3 °C |
| Colombo | 21 °C |
| Cairo | 15 °C |

a) Which city was the warmest?
b) Which city was the coldest?
c) Write the temperatures in ascending order.

**4** Use a number line to show each of these sets of numbers.

a) −5, −2, 0, 5, −3
b) −12, −10, −8, −6, −4
c) All the integers bigger than or equal to −4 but also smaller than or equal to 3.
d) All the integers from −3 to 7.

 **Addition and subtraction of integers**

If a number is positive, we do not always have to write the positive sign (+). For example, $4 = +4$.

If a number is negative, we must always write the negative sign (−) because $-4 \neq 4$.

What happens if there are **two signs** in front of a number?
Here are some rules.

- + (+) can be replaced by +
- − (−) can be replaced by +
- + (−) can be replaced by −
- − (+) can be replaced by −

NOTE: You will learn more about these rules when you study the multiplication of integers later in the unit. Until then, just learn the rules!

To add and subtract any integers, follow these steps.

First look at the signs. If there are **two signs** in front of any number, use the rules above to change them into **one sign**.

Then use a number line to help you to work out the answer.
Remember:

- The **first number** tells us where we **start** on the number line.
- The **next sign** tells us the **direction** in which we must move.
  A + sign tells us to move to the right.
  A − sign tells us to move to the left.
- The **second number** tells us **how many numbers** we must move on the number line.

**Example**

Work out $5 + (-7)$.

$5 + (-7)$ becomes $5 - 7$ because $+ (-)$ can be replaced by −.
We start at the number 5.
The − sign tells us to move to the left.
The 7 tells us to move along 7 numbers.

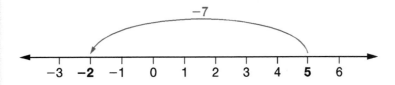

The answer is −2.

If you practise this again and again, you will be able to work out the answer correctly without needing to draw a number line each time, but you should always draw a number line if you are unsure.

**Examples**

Work out these.

a) 7 − 13

We do not have two signs together, so we don't need to change any of them.

We start at 7 and move 13 to the left.

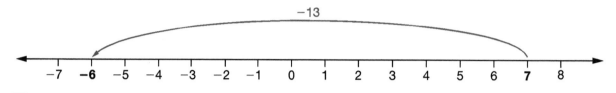

7 − 13 = −6

b) (−11) − 5

We do not have two signs together, so we don't need to change any of them.

We start at −11 and move 5 to the left.

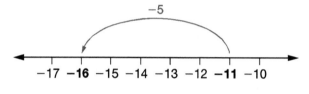

(−11) − 5 = −16

c) (−12) − (−9)

(−12) − (−9) becomes −12 + 9 because − (−) can be replaced by +.

We start at −12 and move 9 to the right.

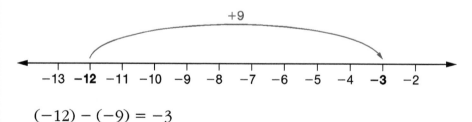

(−12) − (−9) = −3

d) 3 − (−10)

3 − (−10) becomes 3 + 10 because − (−) can be replaced by +.

We start at 3 and move 10 to the right.

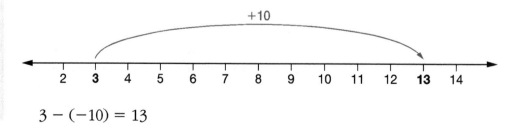

3 − (−10) = 13

Sometimes we need to do several additions and/or subtractions in a row.

**Example**    Calculate $310 + (-145) - (-263) - 42$.

Everywhere there are two signs together, use the rules to change them into one sign.

$$
\begin{aligned}
310 + (-145) - (-263) - 42 &= 310 - 145 + 263 - 42 \\
&= 165 + 263 - 42 \\
&= 428 - 42 \\
&= 386
\end{aligned}
$$

We often use the rules of addition and subtraction in real life.

**Example**    Alec has $50 in his bank account. He writes a cheque for $80. What is the new balance in his account?

We can write the new balance as $50 − $80.
$50 - 80 = -30$
So Alec's account balance is now −$30 (he is $30 **overdrawn**).

## Exercise 11

**1** Use a number line to help you work out the answer to each of these additions.

a) $(-1) + (-4)$        b) $(-2) + (-3)$
c) $(-6) + (-1)$        d) $(-4) + (-2)$
e) $(+3) + (+5)$        f) $(-2) + (-7)$
g) $(-1) + 4$           h) $(-3) + 2$
i) $3 + (-4)$           j) $2 + (-6)$
k) $5 + (-1)$           l) $(-5) + 4$

**2** Work out each of these additions (use a number line if you need to).

a) $(-5) + (-6)$        b) $(-9) + (-7)$
c) $(-3) + (-11)$       d) $(+12) + (+8)$
e) $(-10) + (-11)$      f) $(-4) + (-7) + (-9)$
g) $(-8) + (-17) + (-5) + (-32)$   h) $(-14) + (-5) + (-23) + (-87)$

**3** Work out each of these additions.

a) $(-5) + 13$          b) $(-11) + 19$
c) $14 + (-7)$          d) $23 + (-12)$
e) $(-37) + 22$         f) $(-45) + 19$
g) $25 + (-66)$         h) $74 + (-89)$
i) $101 + (-200)$

4 Add these integers together.
   a) (−7) + (−11) + 9                      b) (−10) + 17 + (−21)
   c) 34 + (−18) + 9                        d) 81 + (−6) + (−62)
   e) 51 + 14 + (−100)                      f) (−27) + 71 + 12
   g) [31 + (−48)] + [(−16) + (−120)]   h) [(−7) + (−14)] + [(−45) + 92]

5 Calculate the correct number of metres (positive or negative) to make each
   of these statements correct. The first has been done for you.

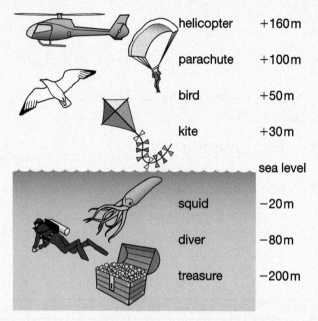

helicopter   +160 m

parachute    +100 m

bird         +50 m

kite         +30 m

sea level

squid        −20 m

diver        −80 m

treasure     −200 m

   a) helicopter + (−60) m = parachute
   b) diver + _____ m = squid
   c) diver + _____ m = treasure
   d) bird + _____ m = squid
   e) parachute + _____ m = treasure
   f) kite + _____ m = squid
   g) helicopter + _____ m = kite
   h) diver + _____ m = helicopter
   i) bird + _____ m = treasure

6 A pizza was taken out of the freezer to thaw before cooking.
   The temperature of the pizza started at −9 °C. After half an hour, the
   temperature had risen by 15°C. What is the new temperature of the pizza?

**7** Use a number line to help you work out the answer to each of these calculations.

a) $8 - (-5)$
b) $(-4) - (-10)$
c) $10 - (+3)$
d) $(-7) - (-6)$
e) $6 - (-1)$
f) $(-5) - (-10)$
g) $(-4) - (+8)$
h) $7 - (-6)$
i) $(-2) - (+9)$
j) $2 - (-9)$
k) $5 - (+5) + 9$
l) $(-10) - (-6) + 4$
m) $(-3) - (-8)$
n) $5 + (-2)$
o) $7 - (+4)$
p) $(-9) - (-5) + (-3)$
q) $7 + (-8) - (+5)$
r) $(-2) - (-7) - 6$
s) $10 + 5 - 8 + 6 - 7$
t) $12 + 8 - 15 + 7 - 20$
u) $30 - 20 + 12 - 50$
v) $6 + 12 - 14 - 4$
w) $37 - 23 - 24 - 25$
x) $12 + 13 + 14 - 20$

**8** Complete these calculations (use a number line if you need to).

a) $5 + (-4) - (-3) + 2 - (-1)$
b) $5 - 4 + (-3) - (-2) + (-1)$
c) $10 - (-11) + (-12) + 13 - (-14)$
d) $(-7) - (-7) + 6 + (-3) + (-9)$
e) $12 + 8 - (-8) + 9 - (-1)$
f) $15 - (-5) + 5 - (-10) + (-20)$
g) $(-5) + (-5) + (-5) + (-5) - (-5)$
h) $5 - (-5) - (-5) - (-5) - (-5)$
i) $1 - (-2) - (-3) - (-4) - (-5)$
j) $(-6) + 7 - 8 + (-9) - (-10)$
k) $(-13) - (-26) + (-39)$
l) $(-44) + 17 - 25 + 13$
m) $[(-7) - 8] - (17 - 7)$
n) $(-356) + 14 - 74$
o) $[(-3) + (-11)] - [17 + (-22)]$
p) $153 - (-230) + (-150) - 99$
q) $(-179) + (-126) - (-169) + 106$
r) $(-16) + 28 - 39 - 47 + 58 - 69 - 73$
s) $21 - 29 + 36 + 45 - 59 - 68 - 79 + 83$
t) $123 - 67 + 44 - 378 + 564 - 88 - 95$

**9** The table shows the temperatures in some cities around the world on Christmas Day.

| City | Temperature on Christmas Day |
|------|------------------------------|
| Edinburgh | −7°C |
| London | 0°C |
| Moscow | −22°C |
| New York | −17°C |
| Rome | 3°C |
| Colombo | 21°C |
| Cairo | 15°C |

What is the difference in temperature if we travel from
a) London to Rome
b) Edinburgh to Rome
c) Moscow to New York
d) Cairo to Colombo
e) Cairo to Moscow?

**10** The temperature inside a freezer is −23°C. The electricity is switched off and the freezer door is opened. After two hours, the temperature has gone up by 8°C. What is the new temperature of the freezer?

 ## Multiplication of integers

We have already learned how to multiply positive numbers together.

We multiply negative numbers in exactly the same way, but we need to know what happens to the signs. Here are some rules.
- positive × positive = positive     (+) × (+) = (+)
- negative × negative = positive     (−) × (−) = (+)
- positive × negative = negative     (+) × (−) = (−)
- negative × positive = negative     (−) × (+) = (−)

NOTE: Always work out the signs first. Then calculate the numbers using short multiplication or long multiplication.

**Examples**     Work out these.

a) $(+7) \times (-5)$
Signs:        $(+) \times (-) = (-)$
Numbers:   $7 \times 5 = 35$
So $(+7) \times (-5) = -35$

b) $(-4) \times (-8)$

Signs: $\quad (-) \times (-) = (+)$

Numbers: $\quad 4 \times 8 = 32$

So $(-4) \times (-8) = 32$

## Division of Integers

We have already learned how to divide positive numbers.

We divide negative numbers in exactly the same way, but we need to know what happens to the signs. The rules are the same as for multiplication.

- positive ÷ positive = positive $\quad (+) \div (+) = (+)$
- negative ÷ negative = positive $\quad (-) \div (-) = (+)$
- positive ÷ negative = negative $\quad (+) \div (-) = (-)$
- negative ÷ positive = negative $\quad (-) \div (+) = (-)$

NOTE: Always work out the signs first. Then calculate the numbers using short division or long division.

**Examples**    Work out these.

a) $(+8) \div (-2)$

Signs: $\quad (+) \div (-) = (-)$

Numbers: $\quad 8 \div 2 = 4$

So $(+8) \div (-2) = -4$

b) $(-18) \div (-3)$

Signs: $\quad (-) \div (-) = (+)$

Numbers: $\quad 18 \times 3 = 6$

So $(-18) \times (-3) = 6$

## Exercise 12

**1** Work out these multiplications.

a) $(+7) \times (+5)$

b) $(-7) \times (+5)$

c) $(-7) \times (-5)$

d) $5 \times (+2)$

e) $(+5) \times (-2)$

f) $(-5) \times (-2)$

g) $(-1) \times (-1)$

h) $8 \times (-3)$

i) $(-8) \times (+3)$

j) $(-5) \times 9$

k) $(-8) \times (-8)$

l) $(-7) \times 6$

m) $(-7) \times (-6)$

n) $8 \times (-10)$

o) $(-8) \times (+10)$

p) $(-4) \times (-8)$

q) $(+5) \times (-2) \times (+2)$

r) $(+4) \times (-3) \times (-5)$

s) $(-3) \times (-2) \times (-5)$

t) $(-5) \times (+3) \times (-4)$

u) $(-5) \times (+3) \times (+4)$

v) $(-5) \times (-4) \times (-5)$

w) $(-4) \times (-3) \times (-9) \times (-2)$

x) $(-3) \times 0 \times (-8) \times (-6) \times 9$

y) $(-7) \times 4 \times (-5) \times (-8) \times 3$

**2** Work out these divisions.

a) $(-8) \div (+2)$        b) $(-8) \div (-2)$

c) $(+20) \div (+4)$      d) $(+20) \div (-4)$

e) $(-20) \div (+4)$      f) $(-20) \div (-4)$

g) $(-18) \div (+3)$      h) $18 \div (-3)$

i) $(-24) \div (-6)$      j) $24 \div (-3)$

k) $(-30) \div (-5)$      l) $(-30) \div (+6)$

m) $(-144) \div 24$      n) $295 \div (-5)$

o) $390 \div (-30)$      p) $(-4864) \div 19$

**3** Work out the answers to these calculations. Remember to follow the correct order of operations.

a) $6 + (-3) \times 4$

b) $(-14) \div 7 + 5$

c) $12 \div (-3) \times 2$

d) $1 - 4 \times (-2)$

e) $5 \times (-4) \div 2$

f) $(-3) + 6 \div (-2)$

g) $(-8) \div (-2) + 3$

h) $(-12) - (-3) \times 2$

i) $(-4) \times 6 + 2$

j) $(-14) \times (-3) - (-9) \div 3$

k) $(-36) \div 9 \times (-16) + (-7) \times 12$

l) $17 \times (-4) - (-5) \times 9 \div (-15)$

m) $47 \times (-4) - (-11) \times (-6) \div 22 \div (-3)$

n) $(-123) \div (-3) + (-87) \times 9 \div (-29)$

# Unit 2 — Factors and multiples

## Key vocabulary

| | | |
|---|---|---|
| common factor | highest common factor (HCF) | prime factor |
| common multiple | index | prime factorisation |
| composite number | index notation | prime number |
| cubed | lowest/least common multiple (LCM) | product |
| divisibility | multiple | squared |
| divisible | power | |
| factor | power number | |

## A  Factors

When numbers are multiplied together, the answer is called the product of the numbers.

We can write any **whole number** as the product of two whole numbers. For example, $3 = 1 \times 3$. Often we can do this in more than one way. For example, $4 = 1 \times 4$ and $4 = 2 \times 2$.

These numbers are called factors of the first number.

Look at the following examples.

$6 = 1 \times 6$ and $6 = 2 \times 3$.
The factors of 6 are 1, 2, 3, and 6.

$12 = 1 \times 12$, $12 = 2 \times 6$ and $12 = 3 \times 4$.
The factors of 12 are 1, 2, 3, 4, 6, and 12.

To find the factors of a number, we need to find all the pairs for that number. 1 and the number itself are the first pair of factors for any number.

**Example**

Find all the factors of 24.

| Start with 1 × the number. | $1 \times 24 = 24$ | 1 and 24 are both factors. |
| Does 2 divide into 24? | $2 \times 12 = 24$ | 2 and 12 are also factors. |
| Does 3 divide into 24? | $3 \times 8 = 24$ | 3 and 8 are also factors. |
| Does 4 divide into 24? | $4 \times 6 = 24$ | 4 and 6 are also factors. |
| Does 5 divide into 24? | No | 5 is not a factor of 24. |
| Does 6 divide into 24? | $6 \times 4 = 24$ | 6 and 4 are both factors. |

We already know that 4 and 6 are factors (the **order** of the factors is not important), so we can stop looking: we have found all the factors of 24: 1, 2, 3, 4, 6, 8, 12 and 24.

## B Multiples

When we multiply one number by another number, the product is called a **multiple** of both these numbers. For example:

$8 \times 2 = 16$   16 is a multiple of 8 **and** a multiple of 2.
$8 \times 3 = 24$   24 is a multiple of 8 **and** a multiple of 3.
$8 \times 4 = 32$   32 is a multiple of 8 **and** a multiple of 4.
… and so on!

To make all the multiples of a number, multiply this number by 1, 2, 3, 4, 5, 6 and so on. For example:

The multiples of 7 are $7 \times 1 = 7$, $7 \times 2 = 14$, $7 \times 3 = 21$, $7 \times 4 = 28$, $7 \times 5 = 35$, $7 \times 6 = 42$, and so on.

We can test whether a number is a multiple of another number by seeing whether it divides **exactly** into it. For example:

21 is a multiple of 7 because $21 \div 7 = 3$ exactly.
23 is not a multiple of 7 because $23 \div 7 = 3$ remainder 2.

1 × a number is called the first multiple, 2 × a number is called the second multiple, 3 × a number is called the third multiple, and so on.

**Examples**

a) Write down the first five multiples of 6.
$1 \times 6 = 6$, $2 \times 6 = 12$, $3 \times 6 = 18$, $4 \times 6 = 24$, $5 \times 6 = 30$
So the first five multiples of 6 are 6, 12, 18, 24 and 30.

**b)** What is the eighth multiple of 9?
   $8 \times 9 = 72$
   So the eighth multiple of 9 is 72.

**c)** The fifth multiple of a number is 30. What is the number?
   You can think of this as $5 \times ? = 30$ or $30 \div 5 = ?$
   $5 \times 6 = 30$ and $30 \div 5 = 6$
   So the number is 6.

## Exercise 1

**1** **a)** Find all the pairs of whole numbers that have a product of 18.
   **b)** Write down all the factors of 18.

**2** **a)** Find all the pairs of whole numbers that have a product of 20.
   **b)** Write down all the factors of 20.

**3** Find all the factors of each of these numbers.
   **a)** 16   **b)** 28   **c)** 36   **d)** 45
   **e)** 48   **f)** 50   **g)** 123   **h)** 144

**4** Find all the factors of each of these numbers.
   **a)** 2   **b)** 3   **c)** 5
   **d)** 7   **e)** 11   **f)** 13
   **g)** Find two more numbers that have only two factors.

**5** Find all the factors of each of these numbers.
   **a)** 4   **b)** 9   **c)** 25   **d)** 49
   **e)** Find two more numbers that have only three factors.

**6** Find all the factors of each of these numbers.
   **a)** 6   **b)** 10   **c)** 14   **d)** 26
   **e)** Find two more numbers that have only four factors.

**7** Is 9 a factor of any of these numbers? Answer 'Yes' or 'No'.
   **a)** 20   **b)** 27   **c)** 31
   **d)** 36   **e)** 54   **f)** 108

**8** Are any of these numbers factors of 48? Answer 'Yes' or 'No'.
   **a)** 1   **b)** 2   **c)** 3   **d)** 4   **e)** 5   **f)** 6
   **g)** 7   **h)** 8   **i)** 9   **j)** 12   **k)** 24   **l)** 48

**9** Write down the first five multiples of each of these numbers.
   **a)** 3   **b)** 7   **c)** 20   **d)** 12

**10** Copy and complete each sentence. Write a word or number in each space to make the sentence correct.

a) The fifth multiple of 4 is _____.

b) The seventh multiple of 10 is _____.

c) The _____ multiple of 6 is 18.

d) The _____ multiple of 8 is 88.

e) The twelfth multiple of _____ is 60.`

f) The fifteenth multiple of _____ is 75.

g) The third multiple of 4 is also the _____ multiple of 6.

h) The fourth multiple of 4 is also the _____ multiple of 8.

i) The tenth multiple of 10 is also the _____ multiple of 20.

j) The fourth multiple of 18 is also the _____ multiple of 24.

**11 a)** Using O for an odd number and E for an even number, complete these multiplication tables.

| ×  | 2 | 3 | 6 | 7 | 9 |
|----|---|---|---|---|---|
| 2  |   |   |   |   |   |
| 3  |   |   |   |   |   |
| 6  |   |   |   |   |   |
| 7  |   |   |   |   |   |
| 9  |   |   |   |   |   |

b) Look at your answers to part a). Can you see a general rule? Use it to complete this table.

| × | O | E |
|---|---|---|
| O |   |   |
| E |   |   |

c) For any number, do you think there will be more even multiples or more odd multiples?

**12** Use what you know about multiples and factors to answer these questions. HINT: Write down the multiples or factors of the numbers given and see which numbers are in both lists.

a) Which multiples of 6 are also factors of 36?

b) Which multiples of 5 are also factors of 120?

c) Which factors of 100 are also multiples of 2?

d) Which factors of 96 are also multiples of 4?

## C Tests of divisibility

We can use these divisibility tests to help us decide quickly whether a number is divisible by some other numbers.

**Divisible by 2**
A whole number that is exactly divisible by 2 is called an **even** number. So, any even number is exactly divisible by 2.

**Divisible by 3**
A number is exactly divisible by 3 if the sum of its digits is divisible by 3. For example:

732 is divisible by 3 because $7 + 3 + 2 = 12$ and 12 is divisible by 3.

**Divisible by 4**
A number is exactly divisible by 4 if the last two digits make a number that is divisible by 4. For example:

35 760 is divisible by 4 because the last two digits are divisible by 4 ($60 \div 4 = 15$).

**Divisible by 5**
A number is exactly divisible by 5 if the last digit is a 5 or a 0.

**Divisible by 9**
A number is exactly divisible by 9 if the sum of the digits is divisible by 9. For example:

67 563 is divisible by 9 because $6 + 7 + 5 + 6 + 3 = 27$ and 27 is divisible by 9.

**Divisible by 10**
A number is exactly divisible by 10 if the last digit is a 0.

**Exercise 2**

**1** Is each of these numbers divisible by i) 2   ii) 3   iii) 5? Answer 'Yes' or 'No'.
 a) 132        b) 225        c) 9997        d) 9018
 e) 53 523     f) 12 358     g) 840 412     h) 132 130

**2** Is each of these numbers divisible by i) 4   ii) 9? Answer 'Yes' or 'No'.
 a) 819        b) 396        c) 1936        d) 9009
 e) 51 450     f) 12 969     g) 156 816     h) 558 944

**3** a) Is the number 5094 divisible by i) 2   ii) 3?
 b) So, is 5094 divisible by 6?
 c) You can use this method to check whether any number is divisible by 6.
    Explain why this works.

**4** a) Is the number 40 054 divisible by i) 2   ii) 7?
 b) So, is 40 054 divisible by 14?
 c) You can use this method to check whether any number is divisible by 14.
    Explain why this works.
 d) Is each of these numbers divisible by 14? Answer 'Yes' or 'No'.
    i) 4237     ii) 6496     iii) 7770     iv) 8514

## D Prime numbers

**Prime numbers** are numbers that have **only two factors**: 1 and the number itself.

The number 1 is not a prime number because it has **only one factor**!
1 is a special number.

The number 2 is the only **even** prime number.

Every number that is > 1 and **not** a prime number, is called a **composite number**. Composite numbers are all numbers that have **more than two factors**.

Is every odd number a prime number?

Is every even number a composite number?

Learn to recognise the prime numbers that are less than 100.

> 2, 3, 5, 7, 11, 13, 17, 19, 23, 29, 31, 37, 41, 43, 47, 53, 59, 61, 67, 71, 73, 79, 83, 89, 97

## E Prime factors

The factors of the number 24 are 1, 2, 3, 4, 6, 8, 12 and 24.
Only some of these factors are prime numbers (2 and 3).

Factors of a number that are prime numbers are called prime factors.

We can write any number as the product of its prime factors using a process called prime factorisation.

**1** If the number is even, divide it by 2. Continue dividing by 2 until the answer is odd.

**2** Is the odd number divisible by 3?
Continue dividing by 3 as many times as you can.

**3** Is the answer after step **2** divisible by 5?
Continue dividing by 5 as many times as you can.

**4** Is the answer after step **3** divisible by 7?
Continue dividing by 7 as many times as you can.

Continue this process with each prime number in turn until your answer is also prime.

**Examples**

Write each of these numbers as the product of its prime factors.

**a)** 120

| 2)120 | 120 is even, so we divide by 2. |
| 2) 60 | 60 is also even so we divide by 2 again. |
| 2) 30 | 30 is also even so we divide by 2 again. |
| 3) 15 | 15 is odd, but is divisible by 3. |
| 5 | 5 is a prime number, so we can stop. |

As the product of its prime factors,
$120 = 2 \times 2 \times 2 \times 3 \times 5$.

**b)** 252

| 2)252 | 252 is even, so we divide by 2. |
| 2)126 | 126 is also even so we divide by 2 again. |
| 3) 63 | 63 is odd, but is divisible by 3. |
| 3) 21 | 21 is also divisible by 3. |
| 7 | 7 is a prime number, so we can stop. |

As the product of its prime factors,
$252 = 2 \times 2 \times 3 \times 3 \times 7$.

**c)** 200

| | |
|---|---|
| 2)200 | 200 is even, so we divide by 2. |
| 2)100 | 100 is also even so we divide by 2 again. |
| 2) 50 | 50 is also even so we divide by 2 again. |
| 5) 25 | 25 is odd. It is not divisible by 3, but is divisible by 5. |
| 5 | 5 is a prime number, so we can stop. |

As the product of its prime factors,
$200 = 2 \times 2 \times 2 \times 5 \times 5$.

## Index notation

In the examples above and on page 47, we wrote
$120 = 2 \times 2 \times 2 \times 3 \times 5$, $252 = 2 \times 2 \times 3 \times 3 \times 7$ and
$200 = 2 \times 2 \times 2 \times 5 \times 5$.

If the **same number** is multiplied more than once, we can write it using index notation: count how many times the same number occurs, and write the total as an index or power number.

Look at the following examples.

| | |
|---|---|
| $5 \times 5 = 5^2$ | We say this as '5 to the power of 2' or '5 squared'. |
| $7 \times 7 \times 7 = 7^3$ | We say this as '7 to the power of 3' or '7 cubed'. |
| $2 \times 2 \times 2 \times 2 = 2^4$ | We say this as '2 to the power of 4'. |
| $3 \times 3 \times 3 \times 3 \times 3 = 3^5$ | We say this as '3 to the power of 5'. |

… and so on.

We sometimes talk about '**powers of 10**'. This means $10^1$ (10), $10^2$ (100), $10^3$ (1000), $10^4$ (10 000), $10^5$ (100 000), …

This works for other numbers as well. So the powers of 2 are $2^1$ (2), $2^2$ (4), $2^3$ (8), $2^4$ (16), $2^5$ (32) … and so on.

We often use index notation when writing a number as a product of its prime factors.

**Examples**

Write each of these numbers as the product of its prime factors.
Use index notation.
**a)** 120
**b)** 252
**c)** 200

We have already found the prime factors for all these numbers, so all we need to do is rewrite them using index notation.

**a)** $120 = 2 \times 2 \times 2 \times 3 \times 5 = 2^3 \times 3 \times 5$

**b)** $252 = 2 \times 2 \times 3 \times 3 \times 7 = 2^2 \times 3^2 \times 7$

**c)** $200 = 2 \times 2 \times 2 \times 5 \times 5 = 2^3 \times 5^2$

NOTE: We will learn more about index or power numbers in Unit 5.

## Exercise 3

**1** Write down all the prime numbers smaller than 100 that are even numbers.

**2** Write down all the prime numbers smaller than 100 that are odd numbers.

**3 a)** Write down the smallest prime number with three digits.
   **b)** Write down the biggest prime number with three digits.

**4** **Twin prime numbers** are two prime numbers that have a difference of 2.
3 and 5 are twin prime numbers ($5 - 3 = 2$).
Write down the next five sets of twin prime numbers.

**5 a)** Write down two prime numbers that add up to an even number.
   **b)** Can one of the numbers be 2?

**6** Answer these questions about what happens when you multiply two prime numbers together.
   **a)** Can the product (answer) be an odd number?
   **b)** Can the product be an even number?
   **c)** Can the product be a prime number?

**7** 37 and 73 are both prime numbers.
They are a special pair because they have the same digits but in opposite order. Find another pair of two-digit prime numbers with reversed digits.

**8** Write each of these numbers as a product of its prime factors.
   **a)** 12    **b)** 20    **c)** 28    **d)** 45    **e)** 66    **f)** 108
   **g)** 250   **h)** 510   **i)** 1089  **j)** 902  **k)** 2028 **l)** 4725

 **9** Here are some numbers, written as the product of their prime factors.
Find each of the numbers.
   **a)** $2^3 \times 5^3$          **b)** $2^6 \times 5^6$
   **c)** $2^2 \times 3^3 \times 5^2$    **d)** $2^4 \times 3^2 \times 5^8$
   **e)** $3^2 \times 5^2 \times 11^4$    **f)** $3^3 \times 5^6 \times 13^3$

## F Highest common factor (HCF)

The prime factors of 20 are $2^2 \times 5$ or $2 \times 2 \times 5$.
The prime factors of 50 are $2 \times 5^2$ or $2 \times 5 \times 5$.

Which prime factors are factors of **both** 20 **and** 50?
Draw a circle around them.

$20 = ⓿2 \times 2 \times ⑤$
$50 = ②\times⑤\times 5$

We can see that 2, 5, and $2 \times 5 = 10$ are all factors of both 20 and 50. We call them the common factors of 20 and 50.

The **biggest** number that is a factor of both 20 and 50 is 10.
We call this the highest common factor or HCF.

To find the **HCF** for any numbers, **multiply** together all of the **common prime factors**.

**Example**  Find the highest common factor (HCF) of 18 and 45.

First write each number as the product of its prime factors and draw a circle around the common factors.

$18 = 2 \times ③ \times ③$
$45 = ③ \times ③ \times 5$

So the HCF of 18 and 45 is $3 \times 3 = 9$.

## G Lowest common multiple (LCM)

The first few multiples of 3 are 3, 6, 9, 12, **15**, 18, 21, 24, 27, **30**, 33, 36, 39, 42, **45**, …

The first few multiples of 5 are 5, 10, **15**, 20, 25, **30**, 35, 40, **45**, 50, …

15, 30, 45, … are multiples of both 3 and 5. We call them the common multiples of 3 and 5.

The **smallest** number that is a multiple of both 3 and 5 is 15.
We call this the lowest common multiple or least common multiple or LCM.

Writing out all the multiples of each number takes a lot of time. To find the **LCM** of any numbers more quickly, look at their prime factors. For each prime factor, find the one with the highest power and multiply these together.

**Examples**

Find the lowest common multiple (LCM) of each of these sets of numbers.

**a)** 15 and 18

First write each number as the product of its prime factors.
For each prime factor, circle the one with the highest power.

$15 = 3 \times \boxed{5}$
$18 = \boxed{2} \times \boxed{3^2}$

There are three different prime factors here: 2, 3 and 5.
The highest power of 2 is 2.
The highest power of 3 is $3^2$.
The highest power of 5 is 5.
So the LCM of 15 and 18 is $2 \times 3^2 \times 5 = 90$.

**b)** 18, 24, and 36.

First write each number as the product of its prime factors.
For each prime factor, circle the one with the highest power.

$18 = 2 \times \boxed{3^2}$
$24 = \boxed{2^3} \times 3$
$36 = 2^2 \times 3^2$

There are two different prime factors here: 2 and 3.
The highest power of 2 is $2^3$.
The highest power of 3 is $3^2$.
So the LCM of 18, 24 and 36 is $2^3 \times 3^2 = 72$.

**Exercise 4**

1 Find the highest common factor (HCF) of each of these sets of numbers.
   a) 12 and 66                        b) 8 and 24
   c) 16 and 18                        d) 20 and 36
   e) 33 and 88                        f) 16, 20 and 28
   g) 15, 39 and 45                    h) 45, 90 and 105
   i) 90 and 126                       j) 1620 and 1728
   k) 8775 and 10 125                  l) 132, 156 and 180
   m) 225, 495 and 810                 n) 972, 1242 and 2538
   o) 96, 144, 136 and 344            p) 140, 168 and 210
   q) $2^3 \times 3^2 \times 5$ and $2^2 \times 3^4 \times 5^3$      r) $2 \times 5^2 \times 7$ and $2^3 \times 3^4 \times 5^3 \times 7^2$

**2** Find the lowest common multiple (LCM) of each of these sets of numbers.

a) 8 and 12

b) 5 and 32

c) 10 and 20

d) 30 and 36

e) 30 and 45

f) 4, 6 and 8

g) 5, 8 and 10

h) 45, 90 and 105

i) 320 and 150

j) 1176 and 1960

k) 153, 204 and 221

l) 98, 140, 210 and 84

m) 168, 336, 784, 1344 and 448

n) $2^3 \times 3^3 \times 5^4$ and $2 \times 3^4 \times 5^3 \times 7$

o) $2^2 \times 5 \times 7$ and $2^3 \times 3^2 \times 5^2 \times 11$

**3** a) Write 108 as a product of its prime factors.

b) Write 162 as a product of its prime factors.

c) What is the HCF of 108 and 162?

d) What is the LCM of 108 and 162?

**4** a) Write 216 as a product of its prime factors.

b) Write 288 as a product of its prime factors.

c) What is the HCF of 216 and 288?

d) What is the LCM of 216 and 288?

**5** The bell at church A rings every 6 minutes. At church B, the bell rings every 9 minutes. Both bells ring together at 9 a.m. When is the next time both bells will ring together?

**6** What is the shortest length of rope that can be cut exactly into 6 m pieces, 8 m pieces or 12 m pieces without anything left over each time?

**7** Find the smallest number that will leave a remainder of 7 when it is divided by each of the numbers 18, 24 and 45.

**8** Find the biggest number that will divide into 197 with a remainder of 5, and also divide into 317 with a remainder of 5.

**9** Find the biggest number that will divide into 11 296 with a remainder of 11, and also divide into 13 528 with a remainder of 23.

**10** I think of a number. When I multiply it by 21, the product will divide exactly by 27, 63 and 108 with no remainder in each case. What is the smallest number I could be thinking of?

# Unit 3  Fractions

| | | |
|---|---|---|
| cancel | half | ordinal numbers |
| decrease | improper fraction | part |
| denominator | increase | proper fraction |
| equal | inequality | quarter |
| equivalent fractions | invert | reduce |
| first | lowest terms | second |
| formula | mixed number | simplify |
| fraction | numerator | third |

## A  What is a fraction?

When we divide a shape into **equal parts**, each part is called a **fraction**.

This shape has **five** equal parts.

Each part is **one out of five** equal parts.

We write this as $\frac{1}{5}$. This is a **fraction**.

In the fraction $\frac{1}{5}$, the number 5 tells us the total number of equal parts that we have.

The number 1 tells us how many of these equal parts we are looking at.

If we now shade two of the parts, we can say that two parts out of the five equal parts are shaded. This fraction, two out of five, is written as $\frac{2}{5}$. We say this as 'two-fifths'.

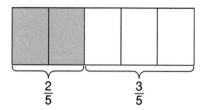

There are three parts out of the five equal parts that are **not** shaded. This fraction, three out of five, is written as $\frac{3}{5}$. We say this as 'three-fifths'.

**Example**

What fraction of the rectangle is shaded?

The rectangle is divided into eight squares.
The squares are all the same size.
**Three out of eight** squares are shaded.
$\frac{3}{8}$ of the rectangle is shaded.

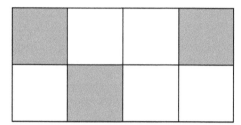

## The parts of a fraction

Each part of a fraction has a special name.

The total number of equal parts is written at the bottom, and is called the denominator.

The number of parts that we are looking at is written at the top, and is called the numerator.

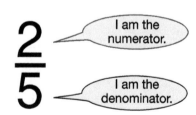

I am the numerator.

I am the denominator.

# Ordinal numbers

We have already learned about the **counting numbers** (0, 1, 2, 3, 4, …). These are the numbers we use to tell us how many of something there are (e.g. there are **35** pupils in the class).

If we need to talk about the **order** of things (e.g. the order in which the pupils are listed in the class), we use another set of numbers called the ordinal numbers. For example:

July is the seventh month of the year.
Nic is the thirteenth pupil in the class.
Thursday is the fourth day of the week.
'Seventh', 'thirteenth' and 'fourth' are ordinal numbers.

The names for most of the ordinal numbers are made by putting 'th' at the end of the name of the counting number (e.g. seven**th**, thirteen**th**, four**th**). However, the ordinal numbers for 1, 2 and 3 have special names.

1  first

2  second

3  third

We use these special names in the ordinal numbers every time the number has a 1, 2 or 3. For example:

21  twenty-first
52  fifty-second
73  seventy-third

There are also some small changes to the names of some other numbers. Here is a list to help you.
The names of the ordinal numbers that do not follow the usual rule exactly, are written in bold.

| Counting number | Symbol | Ordinal number | Symbol |
|---|---|---|---|
| One | 1 | **First** | 1st |
| Two | 2 | **Second** | 2nd |
| Three | 3 | **Third** | 3rd |
| Four | 4 | Fourth | 4th |
| Five | 5 | **Fifth** | 5th |

| Counting number | Symbol | Ordinal number | Symbol |
|---|---|---|---|
| Six | 6 | Sixth | 6th |
| Seven | 7 | Seventh | 7th |
| Eight | 8 | **Eighth** | 8th |
| Nine | 9 | **Ninth** | 9th |
| Ten | 10 | Tenth | 10th |
| Eleven | 11 | Eleventh | 11th |
| Twelve | 12 | **Twelfth** | 12th |
| Thirteen | 13 | Thirteenth | 13th |
| Fourteen | 14 | Fourteenth | 14th |
| Fifteen | 15 | Fifteenth | 15th |
| Sixteen | 16 | Sixteenth | 16th |
| Seventeen | 17 | Seventeenth | 17th |
| Eighteen | 18 | Eighteenth | 18th |
| Nineteen | 19 | Nineteenth | 19th |
| Twenty | 20 | **Twentieth** | 20th |
| Twenty-one | 21 | Twenty-**first** | 21st |
| Twenty-two | 22 | Twenty-**second** | 22nd |
| Twenty-three | 23 | Twenty-**third** | 23rd |
| Twenty-four | 24 | Twenty-fourth | 24th |
| Twenty-five | 25 | Twenty-**fifth** | 25th |
| Twenty-six | 26 | Twenty-sixth | 26th |
| Twenty-seven | 27 | Twenty-seventh | 27th |
| Twenty-eight | 28 | Twenty-**eighth** | 28th |
| Twenty-nine | 29 | Twenty-**ninth** | 29th |
| Thirty | 30 | **Thirtieth** | 30th |
| Forty | 40 | **Fortieth** | 40th |
| Fifty | 50 | **Fiftieth** | 50th |
| Sixty | 60 | **Sixtieth** | 60th |
| Seventy | 70 | **Seventieth** | 70th |
| Eighty | 80 | **Eightieth** | 80th |

| Counting number | Symbol | Ordinal number | Symbol |
|---|---|---|---|
| Ninety | 90 | **Ninetieth** | 90th |
| One hundred | 100 | One hundredth | 100th |
| One hundred and one | 101 | One hundred and **first** | 101st |
| One hundred and two | 102 | One hundred and **second** | 102nd |
| One hundred and three | 103 | One hundred and **third** | 103rd |
| Two hundred | 200 | Two hundredth | 200th |
| Three hundred | 300 | Three hundredth | 300th |
| Four hundred | 400 | Four hundredth | 400th |
| One thousand | 1000 | One thousandth | 1000th |
| Ten thousand | 10 000 | Ten thousandth | 10 000th |
| One million | 1 000 000 | One millionth | 1 000 000th |

We use both these kinds of numbers when we make the names of fractions.

## Naming fractions in English

There are three general rules that we follow when we name fractions.

- We use **counting** numbers for the names of the **numerators**.
- We use **ordinal** numbers for the names of the **denominators**.
- If the numerator is more than one (1), we must add an 's' after the name of the denominator to show there are more than one of them.

Look at the examples at the top of page 58.

There are two exceptions to these rules. $\frac{1}{2}$ and $\frac{1}{4}$ are used a lot and have special names.

- $\frac{1}{2}$ is called 'one-half'
- $\frac{1}{4}$ is called 'one-quarter'

| Fraction | Name | Fraction | Name |
|---|---|---|---|
| $\frac{3}{4}$ | three-quarters | $\frac{1}{30}$ | one-thirtieth |
| $\frac{1}{3}$ | one-third | $\frac{5}{32}$ | five-thirty-seconds |
| $\frac{2}{3}$ | two-thirds | $\frac{2}{21}$ | two-twenty-firsts |
| $\frac{1}{5}$ | one-fifth | $\frac{9}{100}$ | nine-one-hundredths |
| $\frac{3}{5}$ | three-fifths | $\frac{7}{53}$ | seven-fifty-thirds |
| $\frac{1}{21}$ | one-twenty-first | $\frac{5}{12}$ | five-twelfths |

## A second way to name fractions in English

We can see that a fraction is written as one number 'over' another number.

We can use this to give a different name to the fraction. For example, $\frac{5}{8}$ can be named 'five-eighths' **or** 'five **over** eight'.

If the numbers in the fraction are very big, then we **always** use this second way to name the fraction.

Look at the following examples.

| Fraction | Name | Fraction | Name |
|---|---|---|---|
| $\frac{3}{4}$ | three over four | $\frac{1}{30}$ | one over thirty |
| $\frac{1}{3}$ | one over three | $\frac{5}{32}$ | five over thirty-two |
| $\frac{2}{3}$ | two over three | $\frac{2}{21}$ | two over twenty-one |
| $\frac{1}{5}$ | one over five | $\frac{9}{100}$ | nine over one hundred |
| $\frac{3}{5}$ | three over five | $\frac{7}{53}$ | seven over fifty-three |
| $\frac{1}{21}$ | one over twenty-one | $\frac{5}{12}$ | five over twelve |
| $\frac{135}{532}$ | one hundred and thirty-five over five hundred and thirty-two | $\frac{379}{1503}$ | three hundred and seventy-nine over one thousand five hundred and three |

## B Types of fraction

In maths, the words 'out of' mean that we must **divide**.

So every fraction can be written as a division. For example, $\frac{3}{4} = 3 \div 4$. In general, we can say that:

$$\frac{\text{numerator}}{\text{denominator}} \text{ means numerator} \div \text{denominator}$$

We can also write every whole number as a fraction. For example, $5 \div 1 = 5$, so we can write the whole number 5 as the fraction $\frac{5}{1}$. In general, every whole number can be written as the fraction:

$$\frac{\text{whole number}}{1}$$

Each fraction has two parts: the numerator and the denominator. This means that for every fraction, only **one** of these three statements is true.

● numerator < denominator **or**
● numerator = denominator **or**
● numerator > denominator

### Proper fractions

A fraction with a numerator **smaller** than its denominator is called a proper fraction. For example:

$\frac{1}{4}, \frac{2}{3}, \frac{3}{5}, \frac{4}{9}$ and $\frac{9}{13}$ are all proper fractions.

### Improper fractions

A fraction with a numerator **bigger than or equal to** its denominator is called an improper fraction. For example:

$\frac{5}{4}, \frac{4}{3}, \frac{5}{5}, \frac{13}{9}$ and $\frac{17}{3}$ are all improper fractions.

### Mixed numbers

A number with an **integer** part and a **proper fraction** part is called a mixed number. For example:

$1\frac{2}{5}$ and $3\frac{1}{4}$ are both mixed numbers.

We can think of a mixed number as a whole number **+** a fraction.

For example:

$1\frac{2}{5} = 1 + \frac{2}{5}$ (not $1 \times \frac{2}{5}$) and we read it as 'one and two-fifths'.

$3\frac{1}{4} = 3 + \frac{1}{4}$ (not $3 \times \frac{1}{4}$) and we read it as 'three and a quarter'.

## Changing mixed numbers into improper fractions

Look at these mixed numbers. We can use diagrams to represent them.

$1\frac{2}{5}$ can be shown as

| $\frac{1}{5}$ | $\frac{1}{5}$ | $\frac{1}{5}$ | $\frac{1}{5}$ | $\frac{1}{5}$ | $+$ | $\frac{1}{5}$ | $\frac{1}{5}$ |

All together, there are 7 of these parts of $\frac{1}{5}$. As an improper fraction this is written as $\frac{7}{5}$.

$3\frac{1}{4}$ can be shown as

| $\frac{1}{4}$ | $\frac{1}{4}$ | $\frac{1}{4}$ | $\frac{1}{4}$ | $+$ | $\frac{1}{4}$ | $\frac{1}{4}$ | $\frac{1}{4}$ | $\frac{1}{4}$ | $+$ | $\frac{1}{4}$ | $\frac{1}{4}$ | $\frac{1}{4}$ | $\frac{1}{4}$ | $+$ | $\frac{1}{4}$ |

All together, there are 13 of these parts of $\frac{1}{4}$. As an improper fraction this is written as $\frac{13}{4}$.

To change a mixed number into an improper fraction more quickly, we can multiply the denominator of the fraction part by the whole number, and then add the numerator.

$$\text{Improper fraction} = \frac{(\text{denominator} \times \text{whole number}) + \text{numerator}}{\text{denominator}}$$

**Examples**

$$1\frac{2}{5} = \frac{(\text{denominator} \times \text{whole number}) + \text{numerator}}{\text{denominator}}$$

$$= \frac{(5 \times 1) + 2}{5}$$

$$= \frac{5 + 2}{5}$$

$$= \frac{7}{5}$$

$$3\frac{1}{4} = \frac{(\text{denominator} \times \text{whole number}) + \text{numerator}}{\text{denominator}}$$

$$= \frac{(4 \times 3) + 1}{4}$$

$$= \frac{12 + 1}{4}$$

$$= \frac{13}{4}$$

## Changing improper fractions into mixed numbers

This is exactly the opposite of what we have just learned!

To change an improper fraction into a mixed number, we divide the numerator by the denominator to give a whole number and a remainder. The whole number part of the answer is the whole number part of the mixed number. Any remainder from the division makes the numerator of the fraction part.

$$\text{Numerator} \div \text{denominator} = \text{whole number} + \frac{\text{remainder}}{\text{denominator}}$$
$$= \text{mixed number}$$

**Examples**  Change each of these improper fractions into a mixed number.

a) $\frac{23}{7}$

Divide the numerator (23) by the denominator (7).
$23 \div 7 = 3$ remainder 2
3 is the whole number part and 2 is the numerator of the fraction part.
So $\frac{23}{7} = 3 + \frac{2}{7} = 3\frac{2}{7}$

b) $\frac{41}{9}$

$41 \div 9 = 4$ remainder 5
So $\frac{41}{9} = 4 + \frac{5}{9} = 4\frac{5}{9}$

### Exercise 1

**1** Write these fractions using numbers.
For example, one-half $= \frac{1}{2}$.

a) three-quarters
b) one-sixth
c) three-sevenths
d) five-eighths
e) seven-ninths
f) one-tenth
g) six-elevenths
h) eleven-twelfths
i) nine-twentieths

**2** Write these fractions in words.
For example, $\frac{1}{2} =$ one-half.

a) $\frac{1}{4}$
b) $\frac{5}{6}$
c) $\frac{1}{3}$

d) $\frac{3}{8}$
e) $\frac{5}{9}$
f) $\frac{9}{10}$

g) $\frac{6}{7}$
h) $\frac{7}{12}$
i) $\frac{11}{20}$

**3** What fraction of each of these drawings is shaded?

a)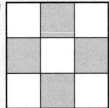

b)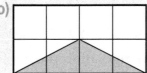

c)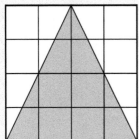

d)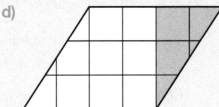

e)

f)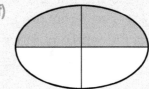

g)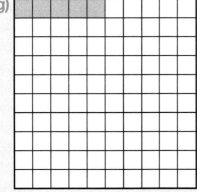

h)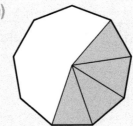

**4** Choose the correct name to describe each of these fractions.
- proper fraction
- improper fraction    or
- mixed number

a) $\frac{3}{2}$          b) $\frac{3}{4}$          c) $\frac{3}{5}$          d) $\frac{12}{3}$          e) $\frac{10}{3}$

f) $1\frac{1}{7}$          g) $\frac{11}{9}$          h) $\frac{110}{99}$          i) $7\frac{1}{10}$

**5** Change each of these improper fractions into a mixed number.

a) $\frac{13}{10}$          b) $\frac{3}{2}$          c) $\frac{17}{8}$          d) $\frac{15}{4}$          e) $\frac{23}{5}$

f) $\frac{34}{7}$          g) $\frac{31}{11}$          h) $\frac{43}{10}$          i) $\frac{28}{9}$

**6** Change each of these mixed numbers into an improper fraction.

a) $2\frac{7}{10}$  b) $1\frac{3}{5}$  c) $5\frac{5}{6}$  d) $3\frac{3}{20}$

e) $4\frac{5}{9}$  f) $7\frac{4}{7}$  g) $11\frac{1}{2}$  h) $4\frac{3}{4}$

**7** Brian cuts a pineapple into eight equal pieces. He eats two of the pieces. What fraction of the pineapple did he eat?

**8** Clare cuts a pizza into twelve equal slices. Her family eats nine of the slices. What fraction of the pizza is left for Clare to eat?

## C Equivalent fractions

Look at the 'fraction wall' below. You will see that each row in the wall is made of different sized 'bricks'. The bricks in each row are all the same size – the same fraction of the whole row.

In the first row there are 2 bricks. Each brick is $\frac{1}{2}$ of the row.

In the second row there are 3 bricks. Each brick is $\frac{1}{3}$ of the row.

… and so on.

| $\frac{1}{2}$ | | | | | | $\frac{1}{2}$ | | | | | |
|---|---|---|---|---|---|---|---|---|---|---|---|
| $\frac{1}{3}$ | | | | $\frac{1}{3}$ | | | | $\frac{1}{3}$ | | | |
| $\frac{1}{4}$ | | | $\frac{1}{4}$ | | | $\frac{1}{4}$ | | | $\frac{1}{4}$ | | |
| $\frac{1}{5}$ | | $\frac{1}{5}$ | | $\frac{1}{5}$ | | $\frac{1}{5}$ | | $\frac{1}{5}$ | | | |
| $\frac{1}{6}$ | | $\frac{1}{6}$ | | $\frac{1}{6}$ | | $\frac{1}{6}$ | | $\frac{1}{6}$ | | $\frac{1}{6}$ | |
| $\frac{1}{7}$ | $\frac{1}{7}$ | | $\frac{1}{7}$ | | $\frac{1}{7}$ | | $\frac{1}{7}$ | | $\frac{1}{7}$ | | $\frac{1}{7}$ |
| $\frac{1}{8}$ | | $\frac{1}{8}$ | | $\frac{1}{8}$ | | $\frac{1}{8}$ | $\frac{1}{8}$ | $\frac{1}{8}$ | | $\frac{1}{8}$ | $\frac{1}{8}$ |
| $\frac{1}{9}$ | $\frac{1}{9}$ | $\frac{1}{9}$ | $\frac{1}{9}$ | $\frac{1}{9}$ | | $\frac{1}{9}$ | $\frac{1}{9}$ | $\frac{1}{9}$ | | $\frac{1}{9}$ | |
| $\frac{1}{10}$ | $\frac{1}{10}$ | $\frac{1}{10}$ | $\frac{1}{10}$ | $\frac{1}{10}$ | | $\frac{1}{10}$ | $\frac{1}{10}$ | $\frac{1}{10}$ | $\frac{1}{10}$ | $\frac{1}{10}$ | |
| $\frac{1}{11}$ | $\frac{1}{11}$ | $\frac{1}{11}$ | $\frac{1}{11}$ | $\frac{1}{11}$ | $\frac{1}{11}$ | $\frac{1}{11}$ | $\frac{1}{11}$ | $\frac{1}{11}$ | $\frac{1}{11}$ | $\frac{1}{11}$ | |
| $\frac{1}{12}$ | $\frac{1}{12}$ | $\frac{1}{12}$ | $\frac{1}{12}$ | $\frac{1}{12}$ | $\frac{1}{12}$ | $\frac{1}{12}$ | $\frac{1}{12}$ | $\frac{1}{12}$ | $\frac{1}{12}$ | $\frac{1}{12}$ | $\frac{1}{12}$ |

Some bricks are shaded. In each case, the shaded bricks make up half the row. So we can see that $\frac{1}{4} + \frac{1}{4} = \frac{2}{4} = \frac{1}{2}$, for example.

If we look at all the shaded bricks we can write

$$\frac{1}{2} = \frac{2}{4} = \frac{3}{6} = \frac{4}{8} = \frac{5}{10} = \frac{6}{12}$$

Even though these fractions look different and have different names, we can see that they are all the same size.

Groups of fractions like this that have different names but are all the same size are called **equivalent fractions**.

Have another look at the fraction wall. Can you find other groups of equivalent fractions?

We can also write $\frac{1}{2} = \frac{2}{4} = \frac{3}{6} = \frac{4}{8} = \frac{5}{10} = \frac{6}{12}$

as $\frac{1}{2} = \frac{1 \times 2}{2 \times 2} = \frac{1 \times 3}{2 \times 3} = \frac{1 \times 4}{2 \times 4} = \frac{1 \times 5}{2 \times 5} = \frac{1 \times 6}{2 \times 6}$

or as $\frac{1}{2} = \frac{2 \div 2}{4 \div 2} = \frac{3 \div 3}{6 \div 3} = \frac{4 \div 4}{8 \div 4} = \frac{5 \div 5}{10 \div 5} = \frac{6 \div 6}{12 \div 6}$

This shows us that if we multiply or divide **both** the numerator **and** the denominator by the same number, the value of the fraction stays the same.

We can use this fact to work out whether two (or more) fractions are equivalent.

Example

Are $\frac{35}{70}$ and $\frac{53}{106}$ equivalent fractions?

Yes, $\frac{35}{70}$ and $\frac{53}{106}$ are both equivalent to $\frac{1}{2}$.

We know this because $\frac{35}{70} = \frac{1 \times 35}{2 \times 35} = \frac{1}{2}$ and $\frac{53}{106} = \frac{1 \times 53}{2 \times 53} = \frac{1}{2}$

or $\frac{35}{70} = \frac{35 \div 35}{70 \div 35} = \frac{1}{2}$ and $\frac{53}{106} = \frac{53 \div 53}{106 \div 53} = \frac{1}{2}$

We can also use this fact to find fractions equivalent to any given fraction – all we need to do is multiply or divide both the numerator and the denominator of the fraction by the same number.

Examples

a) Write down three fractions equivalent to $\frac{5}{7}$.
The numerators must be multiples of 5.
The denominators must be the same multiples of 7.
For example, $\frac{5 \times 2}{7 \times 2} = \frac{10}{14}, \frac{5 \times 3}{7 \times 3} = \frac{15}{21}, \frac{5 \times 4}{7 \times 4} = \frac{20}{28}$ ... and so on.

Three fractions equivalent to $\frac{5}{7}$ are $\frac{10}{14}, \frac{15}{21}$ and $\frac{20}{28}$.

b) The fraction $\frac{2}{3}$ is equivalent to $\frac{?}{12}$. Find the value of the unknown numerator.

The denominator, 3, has been multiplied by 4 to get 12.

To find the unknown numerator, we must also multiply the numerator, 2, by 4.

$$\frac{2 \times 4}{3 \times 4} = \frac{8}{12}$$

So the unknown numerator is $2 \times 4 = 8$

NOTE: When we multiply a number by another number, we make a bigger number. We say that we have **increased** the number.

When we divide a number by another number, we make a smaller number. We say that we have **decreased** or **reduced** the number. Look at these equivalent fractions.

$$\frac{6}{18} = \frac{6 \div 2}{18 \div 2} = \frac{3}{9} = \frac{3 \div 3}{9 \div 3} = \frac{1}{3}$$

When we make an equivalent fraction by dividing the numerator and denominator by the same number, we are 'reducing' (or making smaller) the numerator and denominator. We **reduce the fraction to lower terms**.

We have reduced $\frac{6}{18}$ to the lower terms of $\frac{3}{9}$. We have then reduced $\frac{3}{9}$ some more to the lower terms of $\frac{1}{3}$. We cannot reduce $\frac{1}{3}$ to lower terms because the numbers 1 and 3 have no **common factors** except 1. We have reduced $\frac{6}{18}$ to its **lowest terms**, $\frac{1}{3}$.

We reduce a fraction by dividing by a common factor of the numerator and denominator again and again. We call this process **cancelling**.

We can reduce a fraction to its lowest terms more quickly by dividing both numerator and denominator by their **highest common factor**.

For example, the HCF of 6 and 18 is 6, so $\frac{6}{18} = \frac{6 \div 6}{18 \div 6} = \frac{1}{3}$.

When we reduce a fraction to its lowest terms, we change it into the **simplest** (smallest) fraction that is equivalent to it – we **simplify** the fraction.

When we write a fraction, we should always reduce it to its lowest terms.

NOTE: To check if fractions are equivalent, reduce them to their lowest terms. Equivalent fractions in their lowest terms are exactly the same fraction!

**Examples**

a) Simplify $\frac{24}{30}$.

The HCF of 24 and 30 is 6.

So $\frac{24}{30} = \frac{24 \div 6}{30 \div 6} = \frac{4}{5}$

b) In a class of 28 pupils there are 12 boys.
What fraction of the pupils are boys?
Simplify this fraction to its lowest terms.
There are 12 boys out of 28 students.

The fraction of boys $= \frac{12}{28} = \frac{12 \div 4}{28 \div 4} = \frac{3}{7}$

$\frac{3}{7}$ of the pupils are boys.

c) Write 42 as a fraction of 70.
42 as a fraction of 70 is $\frac{42}{70}$, but this is not in its lowest terms.
We can simplify the fraction by cancelling.

2 is a common factor of 42 and 70 $\qquad \frac{42}{70} = \frac{42 \div 2}{70 \div 2} = \frac{21}{35}$

7 is a common factor of 21 and 35 $\qquad \frac{21}{35} = \frac{21 \div 7}{35 \div 7} = \frac{3}{5}$

3 and 5 have no common factors.
42 as a fraction of 70 is $\frac{3}{5}$.

Another way to cancel is to write both numerator and denominator as products of their **prime factors** and cancel any that appear in both.

For example, $\frac{42}{70} = \frac{\cancel{2} \times 3 \times \cancel{7}}{\cancel{2} \times 5 \times \cancel{7}} = \frac{3}{5}$

**Exercise 2**

**1** Copy and complete each of these calculations, to make a pair of equivalent fractions.

a) $\dfrac{1}{2} = \dfrac{1 \times 9}{2 \times 9} =$

b) $\dfrac{2}{3} = \dfrac{2 \times 8}{3 \times 8} =$

c) $\dfrac{5}{4} = \dfrac{5 \times 6}{4 \times 6} =$

d) $\dfrac{7}{5} = \dfrac{7 \times 7}{5 \times 7} =$

e) $\dfrac{5}{6} = \dfrac{5 \times 3}{6 \times 3} =$

f) $\dfrac{3}{7} = \dfrac{3 \times 4}{7 \times 4} =$

g) $\dfrac{11}{8} = \dfrac{11 \times 2}{8 \times 2} =$

h) $\dfrac{10}{9} = \dfrac{10 \times 10}{9 \times 10} =$

i) $\dfrac{7}{10} = \dfrac{7 \times 9}{10 \times 9} =$

j) $\dfrac{5}{12} = \dfrac{5 \times 3}{12 \times 3} =$

k) $\dfrac{3}{16} = \dfrac{3 \times 11}{16 \times 11} =$

l) $\dfrac{5}{9} = \dfrac{5 \times 5}{9 \times 5} =$

**2** Find the missing number in each of these pairs of equivalent fractions.

a) $\dfrac{1}{3} = \dfrac{\square}{12}$

b) $\dfrac{3}{4} = \dfrac{\square}{12}$

c) $\dfrac{4}{5} = \dfrac{24}{\square}$

d) $\dfrac{5}{6} = \dfrac{25}{\square}$

e) $\dfrac{8}{7} = \dfrac{64}{\square}$

f) $\dfrac{11}{8} = \dfrac{\square}{72}$

g) $\dfrac{4}{13} = \dfrac{\square}{169}$

h) $\dfrac{7}{9} = \dfrac{105}{\square}$

i) $\dfrac{17}{9} = \dfrac{\square}{99}$

j) $\dfrac{30}{100} = \dfrac{3}{\square}$

k) $\dfrac{250}{750} = \dfrac{1}{\square}$

l) $\dfrac{56}{96} = \dfrac{\square}{12}$

**3** Find the missing numbers in each of these sets of equivalent fractions.

a) $\dfrac{2}{9} = \dfrac{\square}{18} = \dfrac{6}{\square}$

b) $\dfrac{7}{10} = \dfrac{\square}{20} = \dfrac{28}{\square}$

c) $\dfrac{12}{11} = \dfrac{36}{\square} = \dfrac{\square}{55}$

d) $\dfrac{17}{12} = \dfrac{34}{\square} = \dfrac{\square}{60}$

e) $\dfrac{5}{6} = \dfrac{\square}{18} = \dfrac{60}{\square}$

f) $\dfrac{3}{\square} = \dfrac{24}{64} = \dfrac{6}{\square}$

g) $\dfrac{2}{3} = \dfrac{8}{\square} = \dfrac{\square}{27} = \dfrac{20}{\square}$

h) $\dfrac{3}{4} = \dfrac{\square}{8} = \dfrac{24}{\square} = \dfrac{21}{\square}$

**4** Reduce each of these fractions to its lowest terms.

a) $\dfrac{16}{24}$

b) $\dfrac{20}{25}$

c) $\dfrac{18}{24}$

d) $\dfrac{14}{49}$

e) $\dfrac{42}{48}$

f) $\dfrac{72}{81}$

g) $\dfrac{16}{40}$

h) $\dfrac{12}{50}$

i) $\dfrac{52}{65}$

j) $\dfrac{75}{90}$

k) $\dfrac{132}{144}$

l) $\dfrac{70}{182}$

m) $\dfrac{26}{39}$

n) $\dfrac{29}{87}$

o) $\dfrac{66}{143}$

p) $\dfrac{84}{300}$

q) $\dfrac{44}{154}$

r) $\dfrac{198}{462}$

s) $\dfrac{525}{1155}$

t) $\dfrac{625}{1000}$

u) $\dfrac{3528}{6552}$

**5** What fraction of each shape is shaded? Write each fraction in its lowest terms.

a) 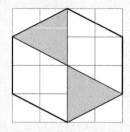  b) 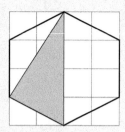  c)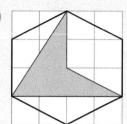

d) 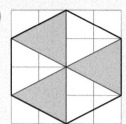  e)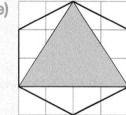

**6** Write each of these mixed numbers in its lowest terms.

a) $1\frac{2}{4}$    b) $2\frac{4}{6}$    c) $3\frac{2}{8}$    d) $4\frac{9}{15}$    e) $5\frac{35}{42}$

f) $6\frac{36}{42}$    g) $7\frac{15}{35}$    h) $8\frac{32}{40}$    i) $9\frac{12}{54}$    j) $10\frac{55}{60}$

**7** Show which of these pairs of fractions are equivalent fractions.

a) $\frac{4}{6}$ and $\frac{10}{15}$    b) $\frac{21}{28}$ and $\frac{12}{16}$    c) $\frac{5}{16}$ and $\frac{35}{102}$

d) $\frac{44}{16}$ and $2\frac{3}{4}$    e) $\frac{20}{48}$ and $\frac{35}{84}$    f) $1\frac{1}{2}$ and $\frac{20}{16}$

g) $\frac{48}{36}$ and $\frac{84}{63}$    h) $\frac{15}{9}$ and $1\frac{2}{3}$

**8** Mr Jones is driving from London to Oxford.

a) The total distance is 50 km. Mr Jones stops for petrol after 35 km.
What fraction of the total distance has he travelled? Write this fraction in its lowest terms.

b) Mr Jones arrives in Oxford 60 minutes after he left London.
The stop for petrol took 12 minutes.
What fraction of the total time is this? Write this fraction in its lowest terms.

**9** The pupils in a class are asked how they travel to school. $\frac{1}{2}$ of the pupils travel by bus. $\frac{1}{3}$ of the pupils travel by motorcycle.
In this class there are more than 20 pupils but less than 30 pupils.

a) How many pupils are in the class?

b) The other pupils all travel by car. What fraction of the class travels by car? Write this fraction in its lowest terms.

## D Comparing fractions

$\frac{1}{5}$ shaded

$\frac{3}{5}$ shaded

If we look at the shaded parts of the shapes, it is easy to see that $\frac{1}{5}$ is smaller than $\frac{3}{5}$. We can also say that $\frac{1}{5}$ is less than $\frac{3}{5}$. We write this using the symbol for 'less than' $\frac{1}{5} < \frac{3}{5}$.

When two equal shapes are divided into the **same number of equal parts**, these parts will be the same size and so the **denominators** of the fractions will be the same. In this case we need to compare **only the numerators** to decide which is the bigger fraction.
For $\frac{1}{5}$ and $\frac{3}{5}$, the numerators are 1 and 3. Because $1 < 3$, we can say that $\frac{1}{5} < \frac{3}{5}$.

We can compare the size of fractions only if the size of all the parts is the same. In other words, the denominators must be the same.

If the denominators are different, we must first rewrite the fractions as equivalent fractions with the same denominator. The easiest way to do this is to find the **lowest common multiple** of the denominators. We can then compare the size of the numerators, as before.

**Examples**

a) Which is bigger, $\frac{2}{3}$ or $\frac{7}{12}$?

The LCM of 3 and 12 is 12.

$\frac{7}{12}$ The denominator is already 12 so we can leave this fraction as it is.

$\frac{2}{3}$ We need to multiply the denominator by 4 to get 12,

so $\frac{2}{3} = \frac{2 \times 4}{3 \times 4} = \frac{8}{12}$

$8 > 7$, so $\frac{8}{12} > \frac{7}{12}$, so $\frac{2}{3} > \frac{7}{12}$

So $\frac{2}{3}$ is bigger.

b) Arrange the fractions $\frac{1}{2}$, $\frac{2}{3}$ and $\frac{5}{8}$ in ascending order.

$\frac{1}{2}$, $\frac{2}{3}$ and $\frac{5}{8}$

The LCM of 2, 3 and 8 is 24.

$\frac{1}{2}$ We need to multiply the denominator by 12 to get 24,

so $\frac{1}{2} = \frac{1 \times 12}{2 \times 12} = \frac{12}{24}$

$\frac{2}{3}$ We need to multiply the denominator by 8 to get 24,

so $\frac{2}{3} = \frac{2 \times 8}{3 \times 8} = \frac{16}{24}$

$\frac{5}{8}$ We need to multiply the denominator by 3 to get 24,

so $\frac{5}{8} = \frac{5 \times 3}{8 \times 3} = \frac{15}{24}$

$12 < 15 < 16$, so $\frac{12}{24} < \frac{15}{24} < \frac{16}{24}$, so $\frac{1}{2} < \frac{5}{8} < \frac{2}{3}$

In ascending order the fractions are $\frac{1}{2}$, $\frac{5}{8}$, $\frac{2}{3}$.

If the fractions are negative, remember that numbers on a number line always get smaller towards the left (see Unit 1).

We can think of a negative fraction as a negative numerator divided by a positive denominator. This is because a fraction = numerator ÷ denominator and negative = negative ÷ positive.

**Example**

Which fraction is bigger, $-\frac{5}{7}$ or $-\frac{2}{3}$?

The LCM of 3 and 7 is 21.

$-\frac{5}{7}$ We need to multiply the denominator by 3 to get 21,

so $-\frac{5}{7} = -\frac{5 \times 3}{7 \times 3} = -\frac{15}{21}$ or $\frac{-15}{21}$

$-\frac{2}{3}$ We need to multiply the denominator by 7 to get 21,

so $-\frac{2}{3} = \frac{2 \times 7}{3 \times 7} = -\frac{14}{21}$ or $\frac{-14}{21}$

$-14 > -15$, so $\frac{-14}{21} > \frac{-15}{21}$, so $-\frac{2}{3} > -\frac{5}{7}$

So $-\frac{2}{3}$ is bigger.

**Exercise 3**

1 For each pair of fractions, find out which one is bigger.

a) $\frac{2}{3}$ or $\frac{3}{4}$

b) $\frac{1}{6}$ or $\frac{2}{11}$

c) $\frac{22}{25}$ or $\frac{7}{8}$

d) $-\frac{7}{10}$ or $-\frac{17}{25}$

e) $\frac{11}{12}$ or $\frac{25}{28}$

f) $-\frac{3}{16}$ or $-\frac{5}{24}$

2 Copy and complete each inequality (statement). Write < or > in each space to make the inequality true.

a) $\frac{3}{4} \square \frac{4}{5}$

b) $-\frac{6}{7} \square -\frac{19}{21}$

c) $\frac{5}{9} \square \frac{13}{24}$

d) $\frac{4}{11} \square \frac{5}{13}$

e) $-\frac{8}{15} \square -\frac{11}{20}$

f) $\frac{11}{18} \square \frac{16}{27}$

3 Write each group of fractions in ascending order.

a) $\frac{2}{3}, \frac{3}{4}$ and $\frac{7}{8}$

b) $-\frac{1}{6}, \frac{3}{16}$ and $\frac{5}{24}$

c) $\frac{11}{12}, \frac{11}{18}$ and $\frac{11}{20}$

d) $\frac{5}{9}, \frac{13}{24}$ and $\frac{16}{27}$

4 Write each group of fractions in descending order.

a) $\frac{3}{4}, \frac{4}{5}$ and $\frac{7}{10}$

b) $-\frac{1}{2}, -\frac{3}{5}$ and $\frac{5}{8}$

c) $\frac{2}{3}, \frac{3}{5}$ and $\frac{4}{7}$

d) $\frac{3}{4}, \frac{5}{6}$ and $\frac{7}{10}$

## E Working with fractions

We can add, subtract, multiply and divide fractions.

### Addition and subtraction of fractions

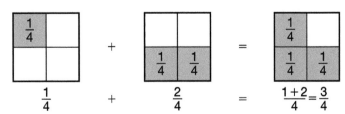

If something is divided into parts that are the same size, we can find the total number of parts by **adding** together the **numerators** of the fractions.

If the fractions have **different denominators**, we must first write them all as **equivalent fractions** with the **same denominators**. Then we can add the numerators as before.

## To add or subtract fractions

1 Find the LCM of all the denominators.

2 Write each of the fractions as an **equivalent fraction** with the LCM as the denominator.

3 Add or subtract the numerators of the new equivalent fractions.

4 If the answer is an **improper fraction**, change it into a **mixed number**.

5 Write any fraction part of the answer in its **lowest terms**.

Sometimes we have to work out an addition or subtraction that includes mixed numbers. We can deal with the mixed numbers in two different ways.
- Add or subtract the whole numbers separately **or**
- Change the mixed numbers into improper fractions.

**Examples**    a) Calculate $\frac{1}{2} + \frac{1}{3}$

$$\begin{aligned}
\frac{1}{2} + \frac{1}{3} &= \frac{1 \times 3}{2 \times 3} + \frac{1 \times 2}{3 \times 2} \\
&= \frac{3}{6} + \frac{2}{6} \\
&= \frac{3 + 2}{6} \\
&= \frac{5}{6}
\end{aligned}$$

The LCM of 2 and 3 is 6.

Remember that we add the numerators only!

b) Calculate $1\frac{7}{10} + \frac{5}{6}$

$$1\frac{7}{10} + \frac{5}{6} = \frac{17}{10} + \frac{5}{6}$$

First write the mixed number as an improper fraction.
The LCM of 10 and 6 is 30.

$$= \frac{17 \times 3}{10 \times 3} + \frac{5 \times 5}{6 \times 5}$$

$$= \frac{51}{30} + \frac{25}{30}$$

$$= \frac{51 + 25}{30}$$

$$= \frac{76}{30}$$

Write the improper fraction as a mixed number.

$$= 2\frac{16}{30}$$

$$= 2\frac{8}{15}$$

Reduce the fraction to its lowest terms.

c) Calculate $3\frac{3}{10} - 1\frac{5}{6}$

**Method 1**

$$3\frac{3}{10} - 1\frac{5}{6} = (3 - 1) + \left(\frac{3}{10} - \frac{5}{6}\right)$$

Deal with the whole numbers separately.

$$= 2 + \left(\frac{3}{10} - \frac{5}{6}\right)$$

Now you can deal with the fraction parts.

$$= 2 + \left(\frac{3 \times 3}{10 \times 3} - \frac{5 \times 5}{6 \times 5}\right)$$

The LCM of 10 and 6 is 30.

$$= 2 + \left(\frac{9}{30} - \frac{25}{30}\right)$$

$$= 2 - \frac{16}{30}$$

$$= \frac{60}{30} - \frac{16}{30}$$

Change the whole number into a fraction.

$$= \frac{60 - 16}{30}$$

$$= \frac{44}{30}$$

$$= 1\frac{14}{30}$$

Change the improper fraction into a mixed number.

$$= 1\frac{7}{15}$$

Reduce the fraction to its lowest terms.

**Method 2**

$$3\frac{3}{10} - 1\frac{5}{6} = \frac{33}{10} - \frac{11}{6}$$

Write the mixed numbers as improper fractions.
The LCM of 10 and 6 is 30.

$$= \frac{33 \times 3}{10 \times 3} - \frac{11 \times 5}{6 \times 5}$$

$$= \frac{99}{30} - \frac{55}{30}$$

$$= \frac{99 - 55}{30}$$

$$= \frac{44}{30}$$

$$= 1\frac{14}{30}$$

Change the improper fraction into a mixed number.
Reduce the fraction to its lowest terms.

$$= 1\frac{7}{15}$$

## Exercise 4

**1** Add these fractions. Write each answer in its lowest terms.

a) $\frac{1}{4} + \frac{1}{8}$     b) $\frac{1}{3} + \frac{1}{4}$     c) $\frac{1}{2} + \frac{1}{5}$

d) $\frac{1}{3} + \frac{1}{5}$     e) $\frac{1}{2} + \frac{1}{7}$     f) $\frac{1}{6} + \frac{1}{8}$

**2** Subtract these fractions. Write each answer in its lowest terms.

a) $\frac{1}{4} - \frac{1}{8}$     b) $\frac{1}{3} - \frac{1}{4}$     c) $\frac{1}{2} - \frac{1}{5}$

d) $\frac{1}{3} - \frac{1}{5}$     e) $\frac{1}{2} - \frac{1}{7}$     f) $\frac{1}{6} - \frac{1}{8}$

**3** Add these fractions. Write each answer in its lowest terms.

a) $\frac{1}{2} + \frac{3}{4}$     b) $\frac{2}{3} + \frac{5}{6}$     c) $\frac{3}{4} + \frac{4}{5}$

d) $\frac{5}{7} + \frac{2}{3}$     e) $\frac{3}{8} + \frac{5}{6}$     f) $\frac{5}{6} + \frac{7}{9}$

**4** Subtract these fractions. Write each answer in its lowest terms.

a) $\frac{5}{8} - \frac{1}{2}$     b) $\frac{13}{15} - \frac{1}{3}$     c) $\frac{5}{6} - \frac{5}{24}$

d) $\frac{7}{9} - \frac{2}{5}$     e) $\frac{7}{8} - \frac{2}{3}$     f) $\frac{13}{16} - \frac{5}{12}$

**5** Add these mixed numbers. Write each answer in its lowest terms.

a) $2\frac{3}{4} + 1\frac{1}{2}$     b) $1\frac{1}{2} + 2\frac{1}{3}$     c) $1\frac{3}{4} + 2\frac{5}{8}$

d) $2\frac{1}{4} + 3\frac{3}{5}$     e) $4\frac{3}{5} + 1\frac{5}{6}$     f) $3\frac{3}{10} + 2\frac{3}{20}$

**6** Subtract these mixed numbers. Write each answer in its lowest terms.

a) $2\frac{1}{2} - 1\frac{2}{5}$

b) $1\frac{2}{3} - 1\frac{1}{4}$

c) $3\frac{3}{8} - 2\frac{3}{4}$

d) $5\frac{2}{5} - 2\frac{1}{10}$

e) $4\frac{3}{10} - 2\frac{5}{8}$

f) $3\frac{3}{16} - 2\frac{5}{24}$

**7** Work out the answers to these calculations. Write each answer in its lowest terms.

a) $\frac{1}{2} + \frac{1}{3} + \frac{1}{6}$

b) $\frac{1}{2} - \frac{1}{4} - \frac{1}{8}$

c) $1\frac{2}{3} + 2\frac{3}{4} + \frac{5}{6}$

d) $9\frac{7}{8} - 6\frac{4}{5} + 3\frac{1}{2}$

e) $\frac{13}{20} + 8\frac{3}{5} - 1\frac{1}{2}$

f) $5\frac{2}{3} - 3\frac{1}{6} - 2\frac{4}{9}$

g) $1\frac{1}{2} - \frac{1}{3} - \frac{3}{4}$

h) $2\frac{1}{3} + 3\frac{1}{6} - \frac{7}{8}$

i) $5\frac{2}{5} - 3\frac{7}{10} + 1\frac{7}{15}$

j) $3\frac{4}{5} + 2\frac{7}{8} + 3\frac{3}{4}$

k) $5\frac{1}{4} - 2\frac{2}{3} - 1\frac{1}{2}$

l) $1\frac{1}{8} + 4\frac{1}{2} + 3\frac{5}{6} + 2\frac{3}{4}$

## Multiplication of fractions

There are 36 pupils in a class. $\frac{2}{3}$ of them are girls. How many girls are in the class?

To answer this question, we need to work out $\frac{2}{3}$ of 36.
In maths the word **of** means the same as **multiply**.
So, $\frac{2}{3}$ of 36 is the same as $\frac{2}{3} \times 36$.

### Multiplying fractions

1 Change any mixed numbers into improper fractions **first**.

2 Simplify by dividing the numerator and denominator by any common factors (if possible). (This is called 'cancelling'.)

3 Multiply together the numbers remaining in the numerators; then multiply together the numbers remaining in the denominators.

4 Simplify your answer.

- If the answer is an improper fraction, change it into a mixed number.
- Write the fraction in its lowest terms.

$$\text{So } \frac{2}{3} \times 36 = 2 \times \frac{\overset{12}{\cancel{36}}}{\underset{1}{\cancel{3}}}$$

$$= 24$$

**Examples**

Calculate these.

a) $2\frac{2}{5} \times 3\frac{1}{4}$

$$2\frac{2}{5} \times 3\frac{1}{4} = \frac{12}{5} \times \frac{13}{4}$$ Change mixed numbers into improper fractions.

$$= \frac{\overset{3}{\cancel{12}}}{5} \times \frac{13}{\underset{1}{\cancel{4}}}$$ Simplify by cancelling.

$$= \frac{39}{5}$$ Multiply the numerators (3 × 13), and multiply the denominators (5 × 1).

$$= 7\frac{4}{5}$$ Change to a mixed number in its lowest terms.

b) $\frac{3}{5} \times \frac{10}{21}$

$$\frac{3}{5} \times \frac{10}{21} = \frac{\overset{1}{\cancel{3}}}{\underset{1}{\cancel{5}}} \times \frac{\overset{2}{\cancel{10}}}{\underset{7}{\cancel{21}}}$$ Simplify by cancelling.

$$= \frac{2}{7}$$ Multiply the numerators (1 × 2), and multiply the denominators (1 × 7).

c) $1\frac{1}{2} \times 1\frac{3}{5}$

$$1\frac{1}{2} \times 1\frac{3}{5} = \frac{3}{2} \times \frac{8}{5}$$ Change mixed numbers into improper fractions.

$$= \frac{3}{\underset{1}{\cancel{2}}} \times \frac{\overset{4}{\cancel{8}}}{5}$$ Simplify by cancelling.

$$= \frac{12}{5}$$ Multiply the numerators (3 × 4), and multiply the denominators (1 × 5).

$$= 2\frac{2}{5}$$ Write as a mixed number in its lowest terms.

 **Exercise 5**

**1** Multiply these fractions. Write each answer in its lowest terms.

a) $\frac{1}{2} \times \frac{1}{3}$  b) $\frac{1}{4} \times \frac{1}{5}$  c) $\frac{3}{5} \times \frac{1}{6}$

d) $\frac{5}{7} \times \frac{1}{3}$  e) $\frac{2}{3} \times \frac{1}{4}$  f) $1\frac{1}{4} \times \frac{1}{5}$

g) $2\frac{2}{5} \times \frac{1}{3}$  h) $2\frac{5}{8} \times \frac{1}{6}$

**2** Multiply these fractions. Write each answer in its lowest terms.

a) $\frac{1}{2} \times \frac{3}{4}$

b) $\frac{3}{4} \times \frac{2}{5}$

c) $\frac{2}{5} \times \frac{5}{6}$

d) $\frac{2}{3} \times \frac{1}{2}$

e) $\frac{3}{10} \times \frac{5}{8}$

f) $\frac{2}{5} \times \frac{1}{7}$

g) $\frac{3}{4} \times \frac{2}{3}$

h) $\frac{3}{10} \times \frac{5}{6}$

i) $\frac{3}{4} \times \frac{2}{9}$

j) $\frac{5}{8} \times \frac{4}{15}$

k) $\frac{5}{21} \times \frac{3}{20}$

l) $\frac{7}{12} \times \frac{4}{21}$

**3** Multiply these fractions and whole numbers. Write each answer in its lowest terms.

a) $\frac{2}{9} \times \frac{3}{5} \times 15$

b) $\frac{2}{15} \times 10 \times \frac{5}{6}$

c) $8 \times \frac{11}{18} \times \frac{13}{22}$

d) $3 \times \frac{9}{20} \times \frac{10}{27}$

**4** Multiply these mixed numbers. Write each answer in its lowest terms.

a) $1\frac{1}{2} \times \frac{3}{4}$

b) $1\frac{1}{2} \times 1\frac{1}{3}$

c) $2\frac{1}{4} \times 1\frac{2}{3}$

d) $2\frac{3}{4} \times 1\frac{2}{5}$

e) $1\frac{1}{2} \times 3\frac{1}{4}$

f) $1\frac{3}{5} \times 1\frac{1}{6}$

g) $6\frac{3}{10} \times 2\frac{2}{9}$

h) $3\frac{3}{4} \times 3\frac{3}{5}$

i) $1\frac{1}{4} \times 2\frac{4}{25}$

j) $2\frac{5}{8} \times 3\frac{1}{3}$

k) $1\frac{1}{2} \times 3\frac{5}{8}$

l) $1\frac{2}{3} \times 4\frac{5}{6}$

**5** Work out the answer to each of these calculations. Write each answer in its lowest terms.

a) $7 \times 2\frac{1}{6} \times 3\frac{1}{7}$

b) $2\frac{1}{2} \times 1\frac{1}{3} \times \frac{1}{10}$

c) $2\frac{3}{4} \times 3\frac{1}{3} \times \frac{7}{22}$

d) $\frac{3}{13} \times 2\frac{5}{6} \times 1\frac{1}{12}$

e) $1\frac{1}{17} \times 1\frac{1}{3} \times 2\frac{5}{6}$

f) $1\frac{1}{3} \times 1\frac{4}{5} \times 5\frac{1}{3}$

**6** Calculate each of these fractions. Write each answer in its lowest terms.

a) $\frac{4}{5}$ of $3\frac{1}{4}$

b) $\frac{2}{3}$ of $5\frac{1}{4}$

c) $\frac{3}{4}$ of $7\frac{1}{5}$

d) $\frac{4}{5}$ of $3\frac{1}{8}$

e) $\frac{3}{8}$ of $3\frac{5}{9}$

f) $\frac{5}{6}$ of $4\frac{2}{7}$

g) $\frac{5}{8}$ of $1\frac{1}{15}$

h) $\frac{3}{10}$ of $4\frac{4}{9}$

**7** Tony eats $\frac{1}{5}$ of a bag of sweets.

a) What fraction of the bag of sweets remains?

Tony divides these remaining sweets into 3 equal parts. Bob gets one of these parts.

b) What fraction of the whole bag of sweets does Bob get?

c) What is the smallest number of sweets that can be in Tony's bag?

## Division of fractions

The diagram shows $\frac{3}{4} \div 2$.

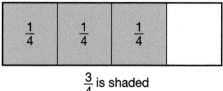

$\frac{3}{4}$ is shaded

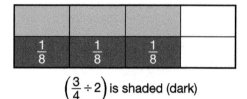

$\left(\frac{3}{4} \div 2\right)$ is shaded (dark)

We can see that $\frac{3}{4} \div 2 = \frac{3}{8}$.

We also know that dividing by two is the same as taking 'half of' it.

$$\begin{aligned} \frac{1}{2} \text{ of } \frac{3}{4} &= \frac{1}{2} \times \frac{3}{4} \\ &= \frac{1 \times 3}{2 \times 4} \\ &= \frac{3}{8} \end{aligned}$$

This shows us that
$$\begin{aligned} \frac{3}{4} \div 2 &= \frac{3}{4} \div \left(\frac{2}{1}\right) \\ &= \frac{3}{4} \times \left(\frac{1}{2}\right) \\ &= \frac{3}{8} \end{aligned}$$

So dividing by a fraction is the same as turning the fraction **after** the ÷ sign upside down and then multiplying by the new fraction. In English, the word for 'turning upside down' is invert.

### Dividing fractions

1 Change any mixed numbers into improper fractions **first**.

2 Invert the fraction **after** the ÷ sign (turn it upside down).

3 Change the ÷ sign into a × sign.

4 Follow the rules for multiplying fractions.

**Examples**    Calculate these.

a) $2\frac{2}{15} \div 1\frac{3}{5}$

$2\frac{2}{15} \div 1\frac{3}{5} = \frac{32}{15} \div \frac{8}{5}$    Change mixed numbers into improper fractions.

$$= \frac{32}{15} \times \frac{5}{8}$$ Change division into multiplication by inverting.

$$= \frac{\overset{4}{\cancel{32}}}{\underset{3}{\cancel{15}}} \times \frac{\overset{1}{\cancel{5}}}{\underset{1}{\cancel{8}}}$$ Simplify by cancelling.

$$= \frac{4}{3}$$ Multiply the numerators, and multiply the denominators.

$$= 1\frac{1}{3}$$ Change to a mixed number in its lowest terms.

b) $\frac{2}{3} \div 5$

$$\frac{2}{3} \div 5 = \frac{2}{3} \times \frac{1}{5}$$ Change division into multiplication by inverting.

$$= \frac{2}{15}$$ Multiply the numerators, and multiply the denominators.

c) $1\frac{3}{5} \div \frac{4}{9}$

$$1\frac{3}{5} \div \frac{4}{9} = \frac{8}{5} \div \frac{4}{9}$$ Change mixed numbers to improper fractions.

$$= \frac{8}{5} \times \frac{9}{4}$$ Change division into multiplication by inverting.

$$= \frac{\overset{2}{\cancel{8}}}{5} \times \frac{9}{\underset{1}{\cancel{4}}}$$ Simplify by cancelling.

$$= \frac{18}{5}$$ Multiply the numerators, and multiply the denominators.

$$= 3\frac{3}{5}$$ Write as a mixed number in its lowest terms.

## Exercise 6

**1** Divide these fractions. Write each answer in its lowest terms.

a) $\frac{1}{2} \div 5$      b) $\frac{3}{4} \div 2$      c) $\frac{2}{5} \div 4$

d) $1\frac{2}{3} \div 5$      e) $1\frac{5}{7} \div 3$      f) $3\frac{1}{9} \div 7$

g) $1\frac{5}{11} \div 4$      h) $2\frac{7}{10} \div 3$

**2** Divide these fractions. Write each answer in its lowest terms.

a) $\frac{1}{2} \div \frac{1}{4}$     b) $\frac{1}{5} \div \frac{1}{2}$     c) $\frac{7}{8} \div \frac{1}{3}$     d) $\frac{2}{3} \div \frac{1}{5}$

e) $2\frac{3}{4} \div \frac{1}{8}$     f) $1\frac{2}{9} \div \frac{1}{6}$     g) $1\frac{12}{19} \div \frac{27}{38}$     h) $6\frac{5}{12} \div 1\frac{5}{6}$

**3** Divide these fractions. Write each answer in its lowest terms.

a) $\frac{2}{3} \div \frac{4}{5}$     b) $\frac{3}{8} \div \frac{2}{3}$     c) $\frac{3}{5} \div \frac{3}{4}$     d) $\frac{2}{5} \div \frac{3}{10}$

e) $\frac{3}{8} \div \frac{9}{16}$     f) $\frac{7}{12} \div \frac{7}{18}$     g) $\frac{4}{9} \div \frac{2}{3}$     h) $\frac{7}{10} \div \frac{3}{5}$

i) $\frac{9}{20} \div \frac{3}{10}$     j) $\frac{21}{25} \div \frac{7}{15}$     k) $1\frac{5}{7} \div 1\frac{3}{5}$     l) $2\frac{3}{4} \div 1\frac{4}{7}$

**4** Divide these mixed numbers. Write each answer in its lowest terms.

a) $2\frac{3}{4} \div 4\frac{1}{8}$     b) $1\frac{1}{2} \div 1\frac{1}{11}$     c) $1\frac{3}{5} \div 1\frac{2}{5}$     d) $6\frac{3}{10} \div 1\frac{7}{20}$

e) $3\frac{3}{4} \div \frac{5}{18}$     f) $1\frac{1}{4} \div 1\frac{9}{16}$     g) $3\frac{1}{5} \div 2\frac{2}{15}$     h) $2\frac{1}{4} \div 1\frac{4}{5}$

i) $4\frac{2}{7} \div 1\frac{7}{8}$     j) $5\frac{2}{5} \div 1\frac{2}{3}$     k) $2\frac{19}{25} \div 1\frac{8}{15}$     l) $9\frac{1}{6} \div 2\frac{3}{4}$

**5** Work out the answer to each of these calculations and write it in its lowest terms.

a) $\frac{1}{2} \div \frac{5}{12} \div \frac{5}{6}$     b) $\frac{1}{2} \times \frac{5}{12} \div \frac{5}{6}$     c) $\frac{1}{2} \div \frac{5}{12} \times \frac{5}{6}$

d) $3\frac{1}{2} \div 2\frac{2}{3} \div 1\frac{3}{4}$     e) $3\frac{1}{2} \times 2\frac{2}{3} \div 1\frac{3}{4}$     f) $3\frac{1}{2} \div 2\frac{2}{3} \times 1\frac{3}{4}$

g) $1\frac{1}{3} \div 2\frac{2}{5} \div \frac{3}{7}$     h) $1\frac{1}{3} \times 2\frac{2}{5} \div \frac{3}{7}$     i) $1\frac{1}{3} \div 2\frac{2}{5} \times \frac{3}{7}$

j) $1\frac{1}{3} \div \frac{2}{5} \div 3\frac{3}{7}$     k) $1\frac{1}{3} \times \frac{2}{5} \div 3\frac{3}{7}$     l) $1\frac{1}{3} \div \frac{2}{5} \times 3\frac{3}{7}$

m) $\frac{1}{3} \div 2\frac{2}{5} \div 1\frac{3}{7}$     n) $\frac{1}{3} \times 2\frac{2}{5} \div 1\frac{3}{7}$     o) $\frac{1}{3} \div 2\frac{2}{5} \times 1\frac{3}{7}$

**6** A shelf is $40\frac{3}{4}$ cm long. Books are $1\frac{9}{10}$ cm wide. Videos are $1\frac{3}{4}$ cm wide.
a) How many books will fit on this shelf?
b) How many videos will fit on this shelf?

**7** $3\frac{3}{4}$ is multiplied by a number and the answer is $2\frac{2}{3}$.
What is the number?

**8** The product of two numbers is 4. One of the numbers is $1\frac{2}{3}$.
What is the other number?

**9** A jug contains $1\frac{3}{4}$ litres of milk. Peter drinks $1\frac{1}{3}$ litres of the milk.
What fraction of all the milk did Peter drink?

## The order of operations with fractions

Do you remember the rules we use to decide the **order** of operations with **integers** (whole numbers)? We use exactly the same rules for calculations involving **fractions**.

1 If there are any **brackets** in the calculation, do the calculation inside the brackets first. You can then remove the brackets. Do not write any brackets if there are none already written!

2 If there are brackets **inside** other brackets, do the calculation in the brackets that are 'most inside' first.

3 If there are **only additions and subtractions** in the calculation, work from left to right with each part in turn.

4 If there are **only multiplications and divisions** in the calculation, work from left to right.

5 If the calculation contains any **combination** of addition and/or subtraction together with multiplication and/or division, then do all the multiplications and/or divisions **before** the additions and/or subtractions.

**Examples**　　Work out these calculations. Write each answer in its lowest terms.

a) $\left(1\frac{1}{2} + \frac{2}{3}\right) \times 1\frac{1}{5}$

$$\left(1\frac{1}{2} + \frac{2}{3}\right) \times 1\frac{1}{5} = \left(\frac{3}{2} + \frac{2}{3}\right) \times \frac{6}{5}$$ 　Change mixed numbers into improper fractions.

$$= \left(\frac{9 + 4}{6}\right) \times \frac{6}{5}$$ 　Do the calculation inside the brackets first.

$$= \frac{13}{6} \times \frac{6}{5}$$

$$= \frac{13}{\cancel{6}_{1}} \times \frac{\cancel{6}^{1}}{5}$$ 　Then do the multiplication.

$$= 2\frac{3}{5}$$ 　Write your answer as a mixed number in its lowest terms.

b) $\left(8\frac{1}{3} \times 5\right) - \left(2\frac{1}{3} \div 3\frac{1}{2}\right)$

Change mixed numbers into improper fractions.

$$\left(8\frac{1}{3} \times 5\right) - \left(2\frac{1}{3} \div 3\frac{1}{2}\right)$$

$$= \left(\frac{25}{3} \times 5\right) - \left(\frac{7}{3} \div \frac{7}{2}\right)$$

$$= \frac{125}{3} - \left(\frac{7}{3} \div \frac{7}{2}\right)$$

Do the calculations inside the brackets first. You might find it easier to concentrate on one at a time, as here.

$$= \frac{125}{3} - \left(\frac{7}{3} \times \frac{2}{7}\right)$$

$$= \frac{125}{3} - \left(\frac{\overset{1}{\cancel{7}}}{3} \times \frac{2}{\underset{1}{\cancel{7}}}\right)$$

$$= \frac{125}{3} - \frac{2}{3}$$

$$= \frac{123}{3}$$

Then do the subtraction.

$$= 41$$

Write your answer in its lowest terms.

## Exercise 7

**1** Use the correct order of operations to work out these calculations.

a) $1 \times \frac{1}{2} - \frac{1}{3} + \frac{1}{4}$  b) $1 - \frac{1}{2} \times \frac{1}{3} + \frac{1}{4}$  c) $1 - \frac{1}{2} + \frac{1}{3} \times \frac{1}{4}$

d) $1 \div \frac{1}{2} + \frac{1}{3} - \frac{1}{4}$  e) $1 + \frac{1}{2} \div \frac{1}{3} - \frac{1}{4}$  f) $1 - \frac{1}{2} + \frac{1}{3} \div \frac{1}{4}$

g) $1 \times \frac{1}{2} - \left(\frac{1}{3} + \frac{1}{4}\right)$  h) $\left(1 - \frac{1}{2}\right) \times \left(\frac{1}{3} + \frac{1}{4}\right)$  i) $1 - \left(\frac{1}{2} + \frac{1}{3}\right) \times \frac{1}{4}$

j) $1 \div \left(\frac{1}{2} + \frac{1}{3}\right) - \frac{1}{4}$  k) $\left(1 - \frac{1}{2}\right) \div \left(\frac{1}{3} - \frac{1}{4}\right)$  l) $1 - \left(\frac{1}{2} + \frac{1}{3}\right) \div \frac{1}{4}$

**2** Use the correct order of operations to work out these calculations.

a) $\frac{1}{2} \div \left(\frac{1}{3} \div \frac{1}{4}\right)$  b) $\frac{1}{2} \div \frac{1}{3} + \frac{1}{4}$  c) $1\frac{1}{10} - \left(\frac{2}{3} \div \frac{5}{6}\right)$

d) $\left(1\frac{1}{10} - \frac{2}{3}\right) \div \frac{5}{6}$  e) $\left(2 - \frac{3}{5}\right) \div \left(3 + \frac{2}{5}\right)$  f) $\left(2 - \frac{1}{2}\right) \div \left(3 + \frac{1}{3}\right)$

g) $1\frac{1}{3} \times \left(4\frac{1}{2} - 1\frac{3}{4}\right)$  h) $1\frac{1}{3} \times 4\frac{1}{2} - 1\frac{3}{4}$  i) $\left(3\frac{1}{2} - 1\frac{2}{3}\right) \div 3\frac{1}{3}$

j) $3\frac{1}{2} - 1\frac{2}{3} \div 3\frac{1}{3}$ 　　　k) $\left(\frac{2}{3} + \frac{1}{3}\right) \times 1\frac{1}{4}$ 　　　l) $\frac{2}{3} + \frac{1}{3} \times 1\frac{1}{4}$

m) $\left(8\frac{1}{3} \div 3\frac{1}{2}\right) - \left(\frac{1}{5} \times 2\frac{1}{3}\right)$ 　n) $8\frac{1}{3} \times \left(3\frac{1}{2} - \frac{1}{5}\right) \div 2\frac{1}{3}$

## Solving word problems with fractions

When you have a word problem to work out:
- Use the dictionary to find out what all the **words** mean (if you don't know them).
- Make sure you understand the whole **story** of the word problem.
- Make sure you understand the **question** and know what you must work out.
- Decide what **calculation** you must do to find the answer to the question.
- When you have the maths answer, make sure that you answer the question using **words**.

**Examples**

a) Many people sleep for 8 hours each day.
What fraction of the day do they sleep?
Number of hours in a day = 24
Number of hours sleeping = 8
Fraction of day sleeping $= \frac{8}{24} = \frac{1}{3}$

Many people spend $\frac{1}{3}$ of the day sleeping.

b) There are 40 pupils in a class. $\frac{3}{5}$ of the pupils are girls.
How many boys are in the class?

**Method 1**
Number of girls $= \frac{3}{5} \times 40$
　　　　　　　$= 24$
Number of boys $= 40 - 24$
　　　　　　　$= 16$
There are 16 boys in the class.

**Method 2**
$\frac{3}{5}$ of the pupils are girls.
Fraction of boys $= 1 - \frac{3}{5}$
　　　　　　$= \frac{2}{5}$
Number of boys $= \frac{2}{5} \times 40$
　　　　　　　$= 16$ boys
There are 16 boys in the class.

c) Mr Jones spends $\frac{1}{3}$ of his salary each month on rent, $\frac{1}{4}$ on food, $\frac{1}{5}$ on transport, $\frac{1}{6}$ on other expenses, and saves what he has left. What fraction of his salary does he save?

Total fraction of salary spent $= \frac{1}{3} + \frac{1}{4} + \frac{1}{5} + \frac{1}{6}$

$$= \frac{20 + 15 + 12 + 10}{60} = \frac{57}{60}$$

Fraction of salary saved $= 1 - \frac{57}{60} = \frac{3}{60}$

$$= \frac{1}{20}$$

d) The petrol tank in a car is two-fifths full.
To fill the tank completely, we need an extra 15 litres.
What is the total number of litres that this petrol tank can hold?
Petrol tank is $\frac{2}{5}$ full.

Extra petrol needed $= 1 - \frac{2}{5} = \frac{3}{5}$

So $\frac{3}{5} = 15$ litres of petrol

and $\frac{1}{5} = 15 \div 3 = 5$ litres.

A full tank is $\frac{5}{5} = 5 \times 5 = 25$ litres.

The tank can hold a total of 25 litres of petrol.

e) Jenny spends $\frac{3}{5}$ of her money. She now has \$1.40 remaining.
How much money did she start with?
Jenny has $1 - \frac{3}{5} = \frac{2}{5}$ of her money left.

So $\frac{2}{5}$ of Jenny's money is \$1.40

and $\frac{1}{5}$ of Jenny's money is $\$1.40 \div 2 = \$0.70$.

All Jenny's money $= \frac{5}{5} = \$0.70 \times 5 = \$3.50$

Jenny started with \$3.50.

f) In a science experiment, we hang a weight on a spring.
The length of the spring increases by $\frac{2}{5}$. The **new** length of the spring is 21 cm.
What was the length of the spring **before** the weight was added?
The starting length of the spring is $\frac{5}{5}$.

Now it is $\frac{2}{5}$ longer.

New length $= \frac{5}{5} + \frac{2}{5} = \frac{7}{5}$

So $\frac{7}{5}$ of the starting length is 21 cm

and $\frac{1}{5}$ of the starting length is $21 \div 7$ cm.

Starting length $= \frac{5}{5} = 3 \times 5 = 15$ cm

The spring was 15 cm long before the weight was added.

**Exercise 8**

1 Work out these fractions and write them in their lowest terms.
   a) 6 hours as a fraction of a day
   b) 4 months as a fraction of a year
   c) 45 seconds as a fraction of $1\frac{1}{4}$ minutes
   d) 24 minutes as a fraction of $2\frac{1}{2}$ hours
   e) 21 minutes as a fraction of $1\frac{1}{2}$ hours
   f) 4 weeks as a fraction of a year

2 Work out these. If necessary, give your answer as a fraction in its lowest terms.
   a) $\frac{3}{8}$ of a day in hours
   b) $\frac{2}{3}$ of a year in months
   c) $\frac{1}{4}$ of $1\frac{1}{4}$ minutes in seconds
   d) $\frac{2}{7}$ of 3 weeks and 3 days, in days
   e) $\frac{3}{5}$ of $1\frac{1}{4}$ hours in seconds
   f) $\frac{3}{4}$ of 2 years in weeks

3 A child pays \$6 to see the circus. A child pays $\frac{3}{4}$ of what an adult pays. How much does an adult pay?

4 A cyclist travels from A to B. He stops after 28 km, which is $\frac{2}{7}$ of the total journey. How long is the total journey from A to B?

5 Ben spends $\frac{5}{8}$ of his pocket money. He now has \$1.20 remaining. How much pocket money did Ben get?

6 Sue has some sweets. She gives $\frac{1}{2}$ of them to her sister and $\frac{1}{3}$ of them to her brother. What fraction of the sweets does she eat?

7 Mr Smith digs his garden. On the first three days he is able to dig $\frac{1}{3}$, $\frac{1}{4}$ and then $\frac{1}{6}$ of it. What fraction of his garden remains for him to dig on the fourth day?

8 The music store has a sale with $\frac{1}{4}$ off all CDs. Robert pays \$9.60 for a CD in the sale. How much money did he save?

9 Mrs McFee has 10 metres of material. She cuts 6 pieces, each $1\frac{1}{4}$ metres long. How much material does she have left?

10 Bob uses $\frac{1}{3}$ of his land to grow durians, $\frac{1}{4}$ for papaya, $\frac{3}{8}$ for guavas, and the remaining 9 hectares for mangoes. How much land does Bob have?

11 Sara gets a pay increase of $\frac{1}{10}$. She now earns \$5.50 for an hour's work. What was her pay for an hour's work **before** the increase?

12 Workers in a factory made 231 chairs in May. This is $\frac{2}{9}$ more than they made in April. How many chairs did they make in April?

## Key vocabulary

| | | |
|---|---|---|
| decimal | decimal whole number | pattern |
| decimal fraction | dividend | place value |
| decimal place | divisor | recurring decimal |
| decimal point | expanded form | repeating decimal |

## A Introduction to decimals

### Place value

In Unit 1 we learned about the **value** of a digit depending on its **place** in a whole number.

Each of the ten digits has a value when it is on its own.

When we combine digits to make a number, the **value** of each digit in the number changes when this digit is in different places in the number. We call this the place value of a digit.

Look at the following examples.

In the number 3978, the digit 8 has a place value of 8.
In the number 5384, the digit 8 has a place value of $8 \times 10 = 80$.
In the number 4853, the digit 8 has a place value of $8 \times 100 = 800$.

If we look at the diagram below, we can see that
- each place gets **10 times bigger** as we move to the **left**
- each place gets **10 times smaller** as we move to the **right**.

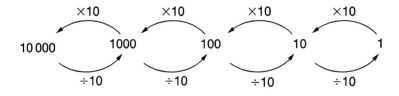

# Decimals

The **decimal** system uses 10 different digits to make all the numbers 0, 1, 2, 3, 4, 5, 6, 7, 8 and 9. We use this system to write whole numbers.

NOTE: Did you know that the word 'deci' in Latin means 10?

We can also use the decimal system as a way to write **fractions**, or parts of a whole number.

To separate the whole number part and the fractional part of a number, we use a **decimal point** (●). This means that the digits for the decimal whole number are on the left-hand side of the decimal point, and the digits for the decimal fraction (the part that is less than 1) are on the right-hand side of the decimal point.

Let's extend the diagram on page 86, so that it includes the **decimal fractions** (numbers that are **smaller than 1**). As before, as we move one place to the right, we divide by 10.

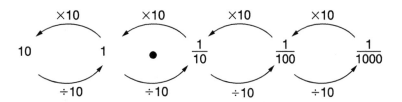

Look at the first place after the decimal point: $1 \div 10 = \frac{1}{10}$

This group contains the fractions: $\frac{1}{10}, \frac{2}{10}, \frac{3}{10}, \frac{4}{10}, \frac{5}{10}, \frac{6}{10}, \frac{7}{10}, \frac{8}{10}$ and $\frac{9}{10}$

We can also write these fractions as $1 \times \frac{1}{10}, 2 \times \frac{1}{10}, 3 \times \frac{1}{10}, 4 \times \frac{1}{10}$, and so on.

So, to find the place value of the digit written in the first place after the decimal point, we multiply by $\frac{1}{10}$ (this is the same as dividing by 10).

Look at the second place after the decimal point: $\frac{1}{10} \div 10$ or $1 \div 100 = \frac{1}{100}$

This group contains the fractions: $\frac{1}{100}, \frac{2}{100}, \frac{3}{100}, \frac{4}{100}, \frac{5}{100}, \frac{6}{100}, \frac{7}{100}, \frac{8}{100}$ and $\frac{9}{100}$

We can also write these fractions as $1 \times \frac{1}{100}, 2 \times \frac{1}{100}, 3 \times \frac{1}{100}, 4 \times \frac{1}{100}$, and so on.

So, to find the place value of the digit written in the second place after the decimal point, we multiply by $\frac{1}{100}$ (this is the same as dividing by 100).

Look at the following examples.

In the number 384.59, the digit 5 has a place value of $5 \times \frac{1}{10} = \frac{5}{10}$

In the number 863.753, the digit 5 has a place value of $5 \times \frac{1}{100} = \frac{5}{100}$

In the number 78.3954, the digit 5 has a place value of $5 \times \frac{1}{1000} = \frac{5}{1000}$

## Understanding decimal numbers

Every digit we write after the decimal point is called a decimal place. For example:

0.7 has one decimal place
1.74 has two decimal places
57.9601 has four decimal places
… and so on.

If there are no whole number digits on the left of the decimal point, we write a '0'. For example:

0.487 is the same as .487. They are both correct, but 0.487 is more usual.

The digits on the right of the decimal point are named as fractions (using the correct power of 10 as the denominator). For example:

$0.487 = \frac{4}{10} + \frac{8}{100} + \frac{7}{1000} = \frac{487}{1000}$

We say 'four-tenths, eight-hundredths, and seven-thousandths' or 'four hundred and eighty-seven-thousandths'.

**Examples**

a) Write these using decimal numbers.
   i) three-tenths and seven-hundredths
   ii) fifteen and six-thousandths

   i) We have no whole numbers.
      After the decimal point we have $\frac{3}{10} + \frac{7}{100}$.
      So the number is 0.37.
   ii) The whole numbers are 15.
      After the decimal point we have $\frac{0}{10} + \frac{0}{100} + \frac{6}{1000}$.
      So the number is 15.006.

b) Write down the place value of the '3' in each of these numbers.
   i) 17.359
   ii) 138.5008
   iii) 4.093 86

i) The 3 is in the tenths column.

$$3 \times \frac{1}{10} = \frac{3}{10}$$

ii) The 3 is in the tens column.

$$3 \times 10 = 30$$

iii) The 3 is in the thousandths columns.

$$3 \times \frac{1}{1000} = \frac{3}{1000}$$

c) How many decimal places does each number in part b) have?

i) There are 3 digits after the decimal point.

17.359 has 3 decimal places.

ii) There are 4 digits after the decimal point.

138.5008 has 4 decimal places.

iii) There are 5 digits after the decimal point.

4.093 86 has 5 decimal places.

Sometimes we need to write a decimal number in expanded form. This means that we write each digit, showing its place value.

**Example**    Write the number 37.802 in expanded form.

$$37.802 = (3 \times 10) + (7 \times 1) + \left(8 \times \tfrac{1}{10}\right) + \left(2 \times \tfrac{1}{1000}\right)$$

**Exercise 1**

1 Write down the place value of the underlined digit in each of these numbers.

a) 6.2<u>3</u>4
b) 123.456 7<u>8</u>9
c) 0.95<u>6</u>7
d) 10.<u>2</u>003
e) 7.947 0<u>5</u>
f) 99.423<u>6</u>
g) <u>6</u>.5702
h) 1<u>0</u>.822
i) 2566.00<u>2</u>
j) 15.042 2<u>4</u>9
k) 76.87<u>6</u> 54<u>3</u> 2
l) <u>2</u>15.363 000 0<u>4</u>

2 Write down the number of decimal places in each of these numbers.

a) 6.234
b) 123.456 789
c) 0.9567
d) 1020.03
e) 7.947 05
f) 99.4236
g) 6570.2
h) 108.22
i) 2566.002
j) 15.042 249
k) 76.876 543 2
l) 215.363 000 04

**3** Copy and complete this table. The first one has been done for you.

| | Decimal number | Expanded form |
|---|---|---|
| | 12.034 | $(1 \times 10) + (2 \times 1) + \left(3 \times \frac{1}{100}\right) + \left(4 \times \frac{1}{1000}\right)$ |
| a) | 567 | |
| b) | | $(6 \times 1000) + (4 \times 10) + (2 \times 1)$ |
| c) | 8.3 | |
| d) | | $(2 \times 1) + \left(7 \times \frac{1}{100}\right)$ |
| e) | 19.02 | |
| f) | | $\left(4 \times \frac{1}{100}\right) + \left(5 \times \frac{1}{1000}\right)$ |
| g) | 0.028 | |
| h) | 90.032 | |
| i) | 9.87 | |
| j) | 1.2345 | |

**4** Write each of these numbers using decimal digits.
  a) six-tenths
  b) two-hundredths
  c) eight-thousandths
  d) two-tenths and six-hundredths
  e) eight-hundredths and one-thousandth
  f) six and nine-tenths
  g) five and two-hundredths
  h) eleven and seven-thousandths
  i) eighteen and six-hundredths
  j) fifty-six, three-tenths and nine-hundredths
  k) two hundred, five-hundredths and seven-thousandths
  l) five hundred, two-tenths and three-thousandths
  m) seventy, four-tenths and five-thousandths
  n) six hundred and thirteen, seven-tenths and eight-ten-thousandths

**5** Write down the next four numbers in each of these patterns.
  a) 0.2, 0.3, 0.4,          b) 0.04, 0.05, 0.06,
  c) 0.83, 0.84, 0.85,       d) 0.21, 0.22, 0.23,
  e) 0.56, 0.58, 0.6,        f) 0.178, 0.18, 0.182,
  g) 0.31, 0.33, 0.35,       h) 0.814, 0.817, 0.82,
  i) 0.945, 0.949, 0.953,    j) 0.1085, 0.109, 0.1095,

**6** What numbers are shown by each of these arrows?

a)
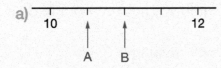

b)
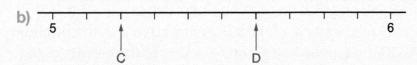

c)
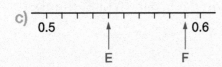

d)
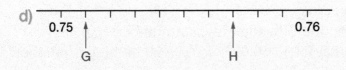

## B Working with decimals

We can compare, order and calculate with decimals, just like integers.

### Comparing decimals

We can tell whether one decimal is bigger or smaller than another decimal by comparing the **place values** of each of the digits in the same places in the numbers. Always start at the **left** of the numbers, and move to the right.

**Example**

Which number is bigger, 45 or 43?

Look at the tens digit in each number. They are both 4; their place value is 40.
Look at the units digit in each number: 5 > 3
This tells us that 45 > 43, so 45 is the bigger number.

We can do exactly the same with numbers that have digits after the decimal point as well.

**Example**

Which number is bigger, 0.63 or 0.69?

Look at the tenths digit in each number. They are both 6; their place value is $\frac{6}{10}$.
Look at the hundredths digit in each number: $\frac{9}{100} > \frac{3}{100}$
This tells us that $0.69 > 0.63$, so 0.69 is the bigger number.

We can do the same with numbers that have a different number of digits – but we must be careful to compare the correct digits.

**Examples**

Look at these pairs of numbers. In each pair, which number is the bigger?
a) 0.03 and 0.003

The first two digits (in the units and tenths columns) in both numbers are zeros.
Look at the hundredths digit in each number: $\frac{3}{100} > 0$
This tells us that $0.03 > 0.003$, so 0.03 is the bigger number.

b) $-6.812$ and $-6.82$

Look at the units digit in each number. The place value of both is $-6$.
Look at the tenths digit in each number. The place value of both is $= -\frac{8}{10}$.

Look at the hundredths digit in each number: $-\frac{1}{100} > -\frac{2}{100}$

This tells us that $-6.812 > -6.82$, so $-6.812$ is the bigger number.

**Exercise 2**

1 Which number is the bigger in each of these pairs?
a) 0.15 and 0.45
b) 0.374 and 0.371
c) 0.405 and 0.4059
d) $-0.27$ and $-0.68$
e) 0.042 and 0.038
f) 0.617 and 0.6173
g) $-0.416$ and $-0.349$
h) 0.08 and 0.008
i) $-0.349$ and $-0.3495$

2 Write $<$ or $>$ in each space to make the inequality (statement) true.
a) 0.65 ☐ 0.47
b) 0.06 ☐ 0.058
c) 29.38 ☐ 29.47
d) 24.923 ☐ 24.913
e) 0.074 ☐ 0.069
f) 7.8 ☐ 7.6
g) 7.045 ☐ 7.036
h) 8.004 ☐ 8.04
i) 0.023 ☐ 0.019
j) 16.23 ☐ 16.45
k) 5.072 ☐ 5.063
l) 6.09 ☐ 6.0009

3  Write each group of decimals in ascending order.
   a) 3.1, 3.01, 3.001, 3.15, 3.2          b) 3.567, 3.657, 3.576, 3.675
   c) 0.1, 0.55, 0.45, 0.5, 0.15           d) 2.93, 2.941, 2.914, 2.939

4  Write each group of decimals in descending order.
   a) 9.87, 8.79, 9.78, 8.97               b) 0.000 15, 0.15, 1.5, 0.015
   c) 2.67, 2.7, 2.599, 2.701              d) 1.993, 1.999, 1.909, 1.939
   e) 2.102, 2.012, 2.201, 2.202

5  Show each group of decimals in ascending order on separate number lines.
   a) 2.7, 2.4, 2.1, 2.55                  b) 3.03, 3.16, 3.12, 3.07
   c) 0.02, 0.08, 0.035, 0.065             d) 1.22, 1.28, 1.31, 1.25, 1.38, 1.345

6  Four boys in Year 7 compare how tall they are. André is 151.54 cm tall,
   Baahir is 152.45 cm tall, Joshua is 152.34 cm tall and Tom is 151.34 cm tall.
   Write down their heights in ascending order.

7  Three pineapples are for sale. Pineapple A weighs 0.909 kg, pineapple B
   weighs 0.99 kg, and pineapple C weighs 0.999 kg.
   a) Which pineapple is the heaviest?
   b) Which pineapple is the lightest?

## Addition and subtraction of decimals

We add and subtract **decimals** in exactly the same way that we add
and subtract **integers**.

1  Write all the numbers in tidy **columns** according to the **place value**
   of the digits. The decimal points should line up under
   each other.

2  Write in zeros to give the same number of decimal places in
   every number.

3  Start the addition or subtraction from the **right**, just as you did
   for whole numbers. Use the same methods for 'carrying' and
   'borrowing' as well.

**Examples** Work out these.

a) $42.6 + 0.75 + 9$

```
    4   2  •  6   0
        0  •  7   5
+       9  •  0   0
    5   2  •  3   5
    1   1
```

b) $17.1 - 8.72$

```
          16        10  1
    0   7  •  1   0
-       8  •  7   2
        8  •  3   8
```

NOTE: Remember that you can use addition to check subtraction.

Sometimes it is easier to work out additions and subtractions in our heads. If we think of a decimal number in expanded form (e.g. $2.5 = 2 + 0.5$), we can add the parts separately.

**Examples** Work out these.

a) $7.8 + 2.5$

$$7.8 + 2.5 = 7.8 + 2 + 0.5$$
$$= 9.8 + 0.5$$
$$= 10.3$$

b) $25.4 - 8.7$

$$25.4 - 8.7 = 25.4 - 8 - 0.7$$
$$= 17.4 - 0.7$$
$$= 16.7$$

### Exercise 3

1 Work out the answer to each of these additions by writing the numbers in columns.
Use subtraction to check your answers.
a) $5.14 + 3.72$
b) $7.065 + 5.384$
c) $11.8 + 5.69$
d) $68.49 + 27.835$
e) $2.86 + 4.7$
f) $3.67 + 10.875$

2 Work out the answer to each of these subtractions by writing the numbers in columns.
Use addition to check your answers.
a) $9.47 - 3.24$
b) $37.6 - 13.28$
c) $45.04 - 20.36$
d) $6.4 - 3.514$
e) $18 - 15.073$
f) $75.1 - 59.173$

3  Work out the answer to each of these calculations in your head.
   a) 2.5 + 8.4                     b) 0.7 + 0.95
   c) 0.36 + 0.54              d) 6.47 + 4.53
   e) 2.7 − 1.5                    f) 1.3 − 0.7
   g) 0.4 − 0.16               h) 15.3 − 6.4

4  Work out the answer to each of these calculations by writing the numbers in columns.
   a) 6.54 + 0.27 + 0.03        b) 79.1 + 7 + 0.23
   c) 10 − 4.78                  d) 9.57 − 4.567
   e) 2.22 + 0.78             f) 9.13 − 7.8936
   g) 5.564 + 0.017 + 10.2      h) 17.1 − 8.82
   i) 9.123 + 0.71 + 6.2 + 1.77   j) 9.123 − 2.85
   k) 37.52 + 15.8 + 2.091      l) 33.7 − 4.81

5  In this question, remember that 1 cent = $0.01. For each part
   i) work out how much I spent
   ii) calculate my change (= money remaining).

   a) I spent 45¢ + 63¢ + 79¢ + $1.43. I paid with a $5.00 note.
   b) I spent $2.47 + $6 + $1.50 + $1.27. I gave the shopkeeper $15.00.
   c) I spent 31¢ + $0.25 + 27¢. I paid with a $10.00 note.
   d) I spend $12 + $3.57 + 67¢. I paid with a $50.00 note.

6  Ann goes to the market. She comes home with 2.5 kg of potatoes, 0.5 kg of grapes, 0.75 kg of fish and 0.6 kg of onions. What is the total weight of her shopping?

7  Fred has a plank of wood 10 m long. He cuts three pieces from the plank, measuring 0.95 m, 1.67 m and 2.5 m. How many metres of wood remain?

8  Simone is 0.15 m shorter than Kai. Kai is 1.7 m tall. How tall is Simone?

9  Mr Monk's monthly salary increased from $1234.92 to $1317.61. What was his pay rise?

10  Swimmer A finishes a race in 51.371 seconds. Swimmer B finishes in 52.090 seconds. What is the difference in the times of these two swimmers?

11  The table shows the monthly rainfall (in mm) for the first six months of 2006 in Chiang Rai, Thailand.

| Jan | Feb | Mar | Apr | May | Jun |
|-----|-----|-----|-----|-----|-----|
| 0.14 | 0.47 | 1.23 | 0.95 | 1.8 | 3.02 |

What was the total rainfall from January to June?

## Multiplying and dividing decimals by 10, 100, 1000, …

When we multiply any decimal by a power of 10, the place value of each digit in the number gets **bigger** by the same power of 10. When we divide any decimal by a power of 10, the place value of each digit in the number gets **smaller** by the same power of 10. We can check this using our knowledge of fractions.

$$0.37 \times 10 = \frac{37}{\overset{10}{\cancel{100}}} \times \frac{\overset{1}{\cancel{10}}}{1} = \frac{37}{10} = 3.7$$

$$0.37 \div 10 = \frac{37}{100} \div \frac{10}{1} = \frac{37}{100} \times \frac{1}{10} = \frac{37}{1000} = 0.037$$

When we **multiply** a decimal by a power of 10, the digits move to the **left** compared with the decimal point; when we **divide** a decimal by a power of 10, the digits move to the **right** compared with the decimal point. You will see that the digits move the same number of places as the number of zeros.

When we multiply a decimal by    **10**    the digits move **1** place to the left.

   **100**    the digits move **2** places to the left.

   **1000**    the digits move **3** places to the left.

… and so on.

When we divide a decimal by    **10**    the digits moves **1** place to the right.

   **100**    the digits move **2** places to the right.

   **1000**    the digits move **3** places to the right.

… and so on.

Look at the following examples.

| | | | 2 | • | 7 | 6 |
|---|---|---|---|---|---|---|
| | | 2 | 7 | • | 6 | |
| | 2 | 7 | 6 | • | | |
| 2 | 7 | 6 | 0 | • | | |

| Multiplication |
|---|
| ← $2.76 \times 10 = 27.6$ |
| ← $2.76 \times 100 = 276$ |
| ← $2.76 \times 1000 = 2760$ |

| 3 | ● | 4 | | | |
|---|---|---|---|---|---|
| 0 | ● | 3 | 4 | | |
| 0 | ● | 0 | 3 | 4 | |
| 0 | ● | 0 | 0 | 3 | 4 |

| Division |
|---|
| ← 3.4 ÷ 10 = 0.34 |
| ← 3.4 ÷ 100 = 0.034 |
| ← 3.4 ÷ 1000 = 0.0034 |

NOTE: Sometimes it is more convenient to think about the **decimal point** moving when we multiply and divide by a power of 10. From the examples above and on page 96, you will see that when we multiply by a power of 10, it seems as if the **decimal point** moves to the **right**. When we divide by a power of 10, it seems as if the **decimal point** moves to the **left**.

## Exercise 4

1 Work out these multiplications.
   a) 25.06 × 10
   b) 25.06 × 100
   c) 25.06 × 1000
   d) 0.93 × 10
   e) 0.93 × 100
   f) 0.93 × 1000
   g) 0.0623 × 10
   h) 0.0623 × 100
   i) 0.0623 × 1000
   j) 9.451 × 10
   k) 9.451 × 100
   l) 9.451 × 1000

2 Work out these divisions.
   a) 37.7 ÷ 10
   b) 37.7 ÷ 100
   c) 37.7 ÷ 1000
   d) 0.27 ÷ 10
   e) 0.27 ÷ 100
   f) 0.27 ÷ 1000
   g) 189.02 ÷ 10
   h) 189.02 ÷ 100
   i) 189.02 ÷ 1000
   j) 9 ÷ 10
   k) 9 ÷ 100
   l) 9 ÷ 1000

3 a) Multiply 0.064 by i) 10   ii) 100   iii) 1000.
   b) Divide 6.4 by i) 10   ii) 100  iii) 1000.

4 An ice cream in Paris costs €1.35.
   a) How much will 10 ice creams cost?
   b) How much will 100 ice creams cost?
   c) How much will 1000 ice creams cost?

5 In London, a pen costs 25p (£1 = 100p). Give your answers to these questions in £.
   a) How much will 10 pens cost?
   b) How much will 100 pens cost?
   c) How much will 1000 pens cost?

6 a) 100 calculators cost $795. How much does one calculator cost?
   b) 1000 pencils cost $120. How much does one pencil cost?
   c) 10 litres of petrol cost $6.69. How much does 1 litre cost?

## Multiplication of decimals

We can use what we already know about multiplying **fractions** to see how to multiply **decimals**.

For example, $15.97 \times 1.9 = \dfrac{1597}{100} \times \dfrac{19}{10}$

$$= \dfrac{1597 \times 19}{1000}$$

We can calculate $1597 \times 19$ using long multiplication (see Unit 1).

```
        1   5   9   7
  ×             1   9
    _____
    1   4   3   7   3
    1   5   9   7   0
    _____
    3   0   3   4   3
```

So $15.97 \times 1.9 = \dfrac{30343}{1000}$

$$= 30.343$$

Notice that the total number of decimal places in the numbers being multiplied ($15.\mathbf{97} \times 1.\mathbf{9} \rightarrow 2 + 1 = 3$) is the same as the total number of decimal places in the answer ($30.\mathbf{343} \rightarrow 3$).

We can check the answer by using whole numbers that are close to the decimals.

15.97 is nearly 16, and 1.9 is nearly 2. So $15.97 \times 1.9 \approx 16 \times 2 \approx 32$. This means that the correct answer is somewhere near 32, and the answer of 30.343 has the right 'size' (the decimal point is in the right place!)

We will learn more about this kind of approximation and estimation in Unit 7.

### How to multiply decimals

**1** Remove the decimal points to make new whole numbers.

**2** Multiply these whole numbers using long multiplication.

**3** Count the total number of decimal places in the given numbers you are multiplying.

**4** Write in the decimal point so that the answer has the same **total number** of decimal places.

**Examples**

Work out these.

**a)** 1.752 × 0.23

|   |   | 1 | 7 | 5 | 2 |
|---|---|---|---|---|---|
| × |   |   |   | 2 | 3 |
|   |   | 5 | 2 | 5 | 6 |
|   | 3 | 5 | 0 | 4 | 0 |
|   | 4 | 0 | 2 | 9 | 6 |

1.752 has 3 decimal places and 0.23 has 2 decimal places, so the answer must have 5 decimal places.

1.752 × 0.23 = 0.402 96

**b)** 4.25 × 0.18

|   |   | 4 | 2 | 5 |
|---|---|---|---|---|
| × |   |   | 1 | 8 |
|   | 3 | 4 | 0 | 0 |
|   | 4 | 2 | 5 | 0 |
|   | 7 | 6 | 5 | 0 |

4.25 has 2 decimal places and 0.18 has 2 decimal places, so the answer must have 4 decimal places.

4.25 × 0.18 = 0.7650

This is usually written as 0.765 because a 0 in the last decimal place has no place value.

Sometimes the answer needs to have more decimal places than there are digits. In these cases, we need to add 0s to the **left** of the digits we do have, until the answer has the correct number of decimal places.

**Example**

Work out 0.4 × 0.2.

4 × 2 = 8

0.4 and 0.2 each have 1 decimal place, so the answer must have 2 decimal places.

We need to add 0s to the left of the 8, until the answer has 2 decimal places.

0.4 × 0.2 = 0.08

**Exercise 5**

**1** Work out each of these multiplications in your head.
  a) $1.7 \times 5$
  b) $1.2 \times 0.3$
  c) $2.6 \times 0.5$
  d) $0.6 \times 0.08$
  e) $0.03 \times 0.2$
  f) $1.2 \times 0.12$
  g) $0.7 \times 0.09$
  h) $0.08 \times 8$

**2** Calculate these products.
  a) $0.01 \times 0.07$
  b) $0.1 \times 0.1$
  c) $0.01 \times 0.01$
  d) $2.5 \times 3.5$
  e) $6.9 \times 4.32$
  f) $28.7 \times 1.9$
  g) $4.12 \times 0.25$
  h) $134 \times 0.73$
  i) $10.7 \times 5.4$
  j) $0.074 \times 0.0024$
  k) $0.48 \times 300$
  l) $8.74 \times 70$
  m) $21.3 \times 0.92$
  n) $2.97 \times 19$
  o) $0.824 \times 6.97$
  p) $91.6 \times 0.098$
  q) $6.73 \times 110$
  r) $24.06 \times 2.003$
  s) $127.6 \times 0.05$
  t) $2 \times 3.14 \times 60$

**3** Work out the cost of these vegetables. (Remember: $100¢ = \$1$)
  a) 0.6 kg of carrots at 35¢ per kilogram.
  b) 4.6 kg of potatoes at 40¢ per kilogram.
  c) 1.2 kg of cabbage at 65¢ per kilogram.

**4** What is the cost of each of these pieces of material?
  a) 7 metres of sheeting at \$1.99 a metre.
  b) 4.5 metres of linen at \$2.24 a metre.
  c) 7.8 metres of satin at \$6.95 a metre.
  d) 3.2 metres of silk at \$8.20 a metre.

**5** €1 can be exchanged for 72.369 Thai baht. What is the value in Thai baht of each of these amounts?
  a) €3.60
  b) €0.60
  c) €54.20

**6** £1 can be exchanged for 2.0452 US dollars. What is the value in US$ of each of these amounts?
  a) £7.50
  b) 75p
  c) £36.50

**7** Here is a list of prices for 1 kilogram of different cheeses in a shop in Sheffield, England.

| Cheddar | Brie | Derby | Stilton |
|---------|------|-------|---------|
| £3.20 | £5.20 | £3.25 | £6.20 |

Work out the cost of each of these pieces of cheese.
a) 0.7 kg of Stilton
b) 1.6 kg of Cheddar
c) 0.8 kg of Derby
d) 0.45 kg of Brie

## Division of decimals

It is difficult to divide by a number that has decimal places: it is much easier to divide by a **whole number**. To change a number with decimal places into a whole number, we must multiply by some power of 10 (10, 100, 1000, ...) depending on the number of decimal places it has. We must also multiply the number that is to be divided, by the **same** power of ten, so that we do not change the value of the calculation.

For example, we can write $2.4 \div 0.4$ as $\frac{2.4}{0.4}$.

We can change this fraction into an equivalent fraction with a whole-number denominator by multiplying both numerator and denominator by 10.

$$\frac{2.4}{0.4} = \frac{2.4 \times 10}{0.4 \times 10} = \frac{24}{4} \qquad \text{So } \frac{2.4}{0.4} = 6$$

NOTE: The number that we **divide by** is called the divisor.
The number that **is divided** is called the dividend.

### How to divide decimals

**1** Multiply the dividing number (divisor) by a power of 10 (10, 100, 1000, ...) so that it becomes a whole number. Multiply the number to be divided (dividend) by the same power of 10.

**2** Complete the division (use long division if you need to).

**3** Be careful to keep the decimal point in the correct place.

**Examples**

Work out these.

a) 2.345 ÷ 0.05

To make the dividing number (0.05) into a whole number, we must multiply by 100.

We must also multiply 2.345 by 100. So the calculation will be 234.5 ÷ 5.

```
     46.9
5)234.5
   20
   ──
   34
   30
   ──
    45
    45
    ──
     0
```

b) 5.196 ÷ 0.24

To make the dividing number (0.24) into a whole number, we must multiply by 100.

We must also multiply 5.196 by 100. So the calculation will be 519.6 ÷ 24.

```
     21.65
24)519.60   ← add zeros if you need
   48
   ──
   39
   24
   ──
   156
   144
   ───
    120
    120
    ───
      0
```

**Exercise 6**

1 Work out these divisions.
   a) 2 ÷ 0.5          b) 3 ÷ 0.2          c) 6 ÷ 0.4
   d) 10 ÷ 2.5         e) 6 ÷ 0.12

2 Work out these divisions.
   a) 7 ÷ 4            b) 8 ÷ 5            c) 1.2 ÷ 0.5
   d) 10.5 ÷ 6         e) 9 ÷ 8

**3** Work out these divisions.

a) 2.46 ÷ 0.2

b) 0.146 ÷ 0.05

c) 2.42 ÷ 0.4

d) 100.1 ÷ 0.07

e) 0.0025 ÷ 0.05

f) 0.05 ÷ 0.004

g) 4.578 ÷ 0.7

h) 0.3 ÷ 0.008

**4** Use long division to work out these divisions.

a) 81.4 ÷ 2.2

b) 15.12 ÷ 2.7

c) 7 ÷ 0.16

d) 11.256 ÷ 0.24

e) 0.1593 ÷ 0.015

f) 16.074 ÷ 0.47

g) 7.647 ÷ 0.25

h) 4.2483 ÷ 1.19

**5** 25 litres of petrol cost $16.70. What is the cost of 1 litre of petrol?

**6** 13 oranges cost $1.43. How much does 1 orange cost?

**7** 1.2 kg of chicken cost $4.32. What is the price of 1 kg of chicken?

**8** 2.5 metres of cloth cost $15.75. What is the cost of 1 metre of cloth?

**9** A piece of wood is 12.73 metres long. I cut this wood into smaller pieces of 0.19 metres long. How many small pieces can I cut?

**10** A jug holds 1.035 litres of water. A small glass holds 0.023 litres. How many small glasses can I fill from the jug?

## Changing decimals into fractions

We already know how to write a decimal in **expanded form**: we write the **place value** for each of the digits.

For example, $0.73 = \left(7 \times \dfrac{1}{10}\right) + \left(3 \times \dfrac{1}{100}\right)$

$$= \frac{7}{10} + \frac{3}{100}$$

If we add these two fractions, we get $\dfrac{7}{10} + \dfrac{3}{100} = \dfrac{70+3}{100}$

$$= \frac{73}{100}$$

So we can say that $0.73 = \dfrac{73}{100}$, or seventy-three-hundredths.

In the same way,

$$0.273 = \left(2 \times \frac{1}{10}\right) + \left(7 \times \frac{1}{100}\right) + \left(3 \times \frac{1}{1000}\right)$$

$$= \frac{2}{10} + \frac{7}{100} + \frac{3}{1000}$$

$$= \frac{200 + 70 + 3}{1000}$$

$$= \frac{273}{1000}, \text{ or two hundred and seventy-three-thousandths}$$

## To write a decimal as a fraction

**1** Write down all the digits that are in decimal places, but do not write the decimal point. These digits will become the **numerator** of the fraction.

**2** The **denominator** will be a power of 10. The number of zeros will be the same as the number of decimal places.

**3** Cancel (if possible) to reduce the fraction to its lowest terms.

**Examples**

Write each of these decimals as a fraction in its lowest terms.

a) 0.3

$$0.3 = \frac{3}{10}$$

b) 0.6

$$0.6 = \frac{6}{10}$$

$$= \frac{3}{5}$$

c) 0.45

$$0.45 = \frac{45}{100}$$

$$= \frac{9}{20}$$

d) 0.037

$$0.037 = \frac{37}{1000}$$

e) 1.5

$$1.5 = 1 + 0.5$$

$$= 1 + \frac{5}{10}$$

$$= 1 + \frac{1}{2}$$

$$= 1\frac{1}{2}$$

f) 31.03

$$31.03 = 31 + 0.03$$

$$= 31 + \frac{3}{100}$$

$$= 31\frac{3}{100}$$

g) 6.75

$$6.75 = 6 + 0.75$$

$$= 6 + \frac{75}{100}$$

$$= 6 + \frac{3}{4}$$

$$= 6\frac{3}{4}$$

h) 0.8

$$0.8 = \frac{8}{10}$$

$$= \frac{4}{5}$$

i) 7.25

$$7.25 = 7 + 0.25$$

$$= 7 + \frac{25}{100}$$

$$= 7 + \frac{1}{4}$$

$$= 7\frac{1}{4}$$

## Changing fractions into decimals

If the denominator of the fraction is 10, 100, 1000, or another power of 10, then it is easy to write these fractions as decimals. The decimal will have the same number of decimal places as there are zeros in the denominator. For example:

$\frac{7}{10} = 0.7$, $\frac{7}{100} = 0.07$, $\frac{27}{1000} = 0.027$ and so on …

We also know that any fraction means numerator ÷ denominator. For example, $\frac{2}{5} = 2 \div 5$. So to change the fraction into a decimal, we must simply carry out this division. For example:

$$
\begin{array}{r}
0.4 \\
5\overline{)2.0} \\
\underline{2\,0} \\
0
\end{array}
$$
← Add zeros when necessary. Remember the decimal point!

So $\frac{2}{5} = 0.4$

### To write a fraction as a decimal

Divide the numerator of the fraction by the denominator.

**Examples**

Write each of these fractions as a decimal.

a) $\frac{3}{8}$

$\frac{3}{8} = 3 \div 8$

$$
\begin{array}{r}
0.375 \\
8\overline{)3.000} \\
\underline{2\,4} \\
60 \\
\underline{56} \\
40 \\
\underline{40} \\
0
\end{array}
$$

$\frac{3}{8} = 0.375$

b) $2\frac{3}{4}$

**Method 1**

First write the mixed number as an improper fraction.

$2\frac{3}{4} = \frac{11}{4} = 11 \div 4$

$$
\begin{array}{r}
2.75 \\
4{\overline{\smash{)}}}11.00 \\
\underline{8\phantom{.00}} \\
30\phantom{.0} \\
\underline{28\phantom{.0}} \\
20 \\
\underline{20} \\
0
\end{array}
$$

$2\frac{3}{4} = 2.75$

**Method 2**

Leave the whole number part, and just change the fraction part to a decimal.

$2\frac{3}{4} = 2 + \frac{3}{4} = 2 + (3 \div 4)$

$$
\begin{array}{r}
0.75 \\
4{\overline{\smash{)}}}3.00 \\
\underline{28\phantom{.0}} \\
20 \\
\underline{20} \\
0
\end{array}
$$

$2\frac{3}{4} = 2 + 0.75 = 2.75$

## Exercise 7

1 Write each of these decimals as a fraction in its lowest terms.

| | | | |
|---|---|---|---|
| a) 0.3 | b) 0.7 | c) 0.07 | d) 0.09 |
| e) 0.009 | f) 0.003 | g) 0.5 | h) 0.6 |
| i) 0.12 | j) 0.25 | k) 0.72 | l) 0.36 |
| m) 0.075 | n) 0.045 | o) 0.48 | p) 0.18 |

**2** Write each of these fractions as a decimal.

a) $\frac{19}{100}$    b) $\frac{23}{1000}$    c) $\frac{1}{5}$    d) $\frac{4}{5}$

e) $\frac{7}{4}$    f) $\frac{3}{4}$    g) $\frac{5}{8}$    h) $\frac{1}{8}$

i) $\frac{7}{20}$    j) $\frac{19}{20}$

**3** Write each of these decimals as a mixed number in its lowest terms.

a) 1.6    b) 1.4    c) 5.002    d) 4.005

e) 12.15    f) 24.35    g) 20.024    h) 50.084

**4** Write each of these mixed numbers as a decimal.

a) $2\frac{3}{10}$    b) $5\frac{7}{10}$    c) $1\frac{2}{25}$    d) $2\frac{1}{25}$

e) $6\frac{1}{4}$    f) $4\frac{3}{4}$    g) $3\frac{3}{8}$    h) $7\frac{7}{8}$

i) $1\frac{5}{16}$    j) $3\frac{3}{16}$

## C Recurring decimals

Sometimes, when you change a fraction into a decimal by dividing the numerator by the denominator, the digits after the decimal point make a pattern of digits that **recurs** (**repeats** again and again). These decimals are called recurring decimals (or sometimes repeating decimals).

We use the symbol $\overset{\bullet}{}$ to show that a digit recurs.

If **the same digit** recurs, we can write this digit only once and put a single dot above it. For example:

$\frac{1}{3} = 0.333\,333\,333\ldots = 0.\dot{3}$

$\frac{5}{6} = 0.833\,333\,333\ldots = 0.8\dot{3}$

If **more than one digit** recurs, we write the recurring pattern of digits only once, and put a dot above the first digit in the pattern and a dot above the last digit in the pattern.

Look at the following examples.

$$\frac{4}{11} = 0.363\,636\,363\ldots = 0.\overset{\bullet\bullet}{3}6$$

$$\frac{8}{37} = 0.216\,216\,216\ldots = 0.\overset{\bullet\;\;\bullet}{2}1\overset{}{6}$$

$$\frac{2}{7} = 0.285\,714\,285\,714\ldots = 0.\overset{\bullet}{2}857\overset{}{1}\overset{\bullet}{4}$$

NOTE: Sometimes, in other books, you might see the symbol ¯ used to show recurring digits instead.

You can check these examples and those on page 107 by dividing the numerator by the denominator.

Every recurring decimal can be written as an exact fraction. We will learn how to change recurring decimals into fractions in Coursebook 2.

## Exercise 8

1 Rewrite each of these decimals, using dots ( • ) to show the recurring digits.
   a) 0.777 777 …
   b) 0.111 111 …
   c) 0.363 636 …
   d) 0.828 282 …
   e) 0.135 135 …
   f) 0.216 216 …
   g) 0.166 666 …
   h) 0.861 111 …
   i) 0.772 727 272 …
   j) 0.123 423 4 …
   k) 0.585 714 285 714 2 …

2 Write each of these fractions as recurring decimals, using dots to show the recurring digits.
   a) $\frac{8}{9}$
   b) $\frac{4}{9}$
   c) $\frac{17}{33}$
   d) $\frac{8}{11}$
   e) $\frac{1}{7}$
   f) $\frac{6}{7}$
   g) $\frac{1}{30}$
   h) $\frac{7}{15}$
   i) $\frac{5}{6}$
   j) $\frac{17}{22}$

# Unit 5 Power numbers

 ## Index form

We have already learned that if the same number is multiplied more than once (e.g. $2 \times 2 \times 2 \times 2 \times 2$), we can write it using index notation (see Unit 2).

To write a number using index notation, count how many times the same number occurs, and write the total as an index or power number. For example:

| Number | Index form | Read as | Value |
|---|---|---|---|
| $2 \times 2 \times 2 \times 2 \times 2$ | $2^5$ | 2 to the power of 5 | 32 |
| $3 \times 3 \times 3 \times 3$ | $3^4$ | 3 to the power of 4 | 81 |
| $5 \times 5 \times 5 \times 5 \times 5 \times 5 \times 5$ | $5^7$ | 5 to the power of 7 | 78 125 |
| $4 \times 4 \times 4 \times 4 \times 4$ | $4^5$ | 4 to the power of 5 | 1024 |
| $6 \times 6 \times 6$ | $6^3$ | 6 to the power of 3 | 216 |
| $2.1 \times 2.1 \times 2.1 \times 2.1$ | $(2.1)^4$ | 2.1 to the power of 4 | 19.4481 |

NOTE: 'Index notation', 'index form', 'power numbers' and 'power form' all mean the same thing. The plural of 'index' is 'indices'.

Look at the numbers $4^3$, $6^5$ and $3^8$. They are all made up of two numbers. We already know that the small, raised number is called the **power** or **index**. The other number is called the base of the power number.

There are two special power numbers, 0 and 1.

**0**     $2^0 = 1$,   $3^0 = 1$,   $4^0 = 1$,   $5^0 = 1$, …

Any number to the power of zero is equal to 1.

**1**     $2^1 = 2$,   $3^1 = 3$,   $4^1 = 4$,   $5^1 = 5$, …

Any number to the power of 1, is equal to that same number.

## Exercise 1

**1** Write each of these numbers using index form.
   a) $4 \times 4 \times 4 \times 4$
   b) $3 \times 3 \times 3 \times 3 \times 3 \times 3 \times 3 \times 3$
   c) $8 \times 8 \times 8 \times 8 \times 8 \times 8 \times 8$
   d) $0.3 \times 0.3 \times 0.3$
   e) $1.6 \times 1.6 \times 1.6 \times 1.6 \times 1.6$
   f) $12 \times 12 \times 12 \times 12 \times 12 \times 12 \times 12$

**2** Copy and complete this table.

| Number | Index form | Value |
|---|---|---|
| $10 \times 10 \times 10 \times 10 \times 10 \times 10$ | $10^6$ | 1 000 000 |
| $10 \times 10 \times 10 \times 10 \times 10$ | $10^5$ | |
| | $10^4$ | |
| $10 \times 10 \times 10$ | | |
| | | 100 |
| 10 | | |
| | | 1 |

**3** Copy and complete these tables.

| Number | Index form | Value |
|---|---|---|
| $5 \times 5 \times 5 \times 5 \times 5 \times 5$ | $5^6$ | 15 625 |
| $5 \times 5 \times 5 \times 5 \times 5$ | $5^5$ | |
| | $5^4$ | |
| $5 \times 5 \times 5$ | | |
| | | 25 |
| 5 | | |
| | | 1 |

| Number | Index form | Value |
|---|---|---|
| $4 \times 4 \times 4 \times 4 \times 4 \times 4$ | $4^6$ | 4096 |
| $4 \times 4 \times 4 \times 4 \times 4$ | $4^5$ | |
| | $4^4$ | |
| $4 \times 4 \times 4$ | | |
| | | 16 |
| 4 | | |
| | | 1 |

**4** Work out the value of each of these power numbers.

    a) $2^3$                               b) $6^2$
    c) $3^4$                               d) $12^2$
    e) $5^4$                               f) $10^7$

**5** Write each of these numbers using index form.
Example: $2 \times 5 \times 5 \times 7 \times 7 \times 7 = 2 \times 5^2 \times 7^3$

    a) $2 \times 2 \times 3 \times 3$                   b) $2 \times 2 \times 2 \times 3 \times 3 \times 5$
    c) $2 \times 3 \times 3 \times 3 \times 5$            d) $2 \times 3 \times 5 \times 5$
    e) $2 \times 2 \times 2 \times 2 \times 3 \times 5 \times 5$     f) $3 \times 3 \times 3 \times 3 \times 5 \times 5 \times 7$

## B   Multiplying and dividing numbers with powers

Look at these multiplications.

$$6^5 \times 6^4 = (6 \times 6 \times 6 \times 6 \times 6) \times (6 \times 6 \times 6 \times 6)$$
$$= 6 \times 6 \times 6 \times 6 \times 6 \times 6 \times 6 \times 6 \times 6$$
$$= 6^9$$
$$3^2 \times 3^6 = (3 \times 3) \times (3 \times 3 \times 3 \times 3 \times 3 \times 3)$$
$$= 3 \times 3 \times 3 \times 3 \times 3 \times 3 \times 3 \times 3$$
$$= 3^8$$

So $6^5 \times 6^4 = 6^9$ and $3^2 \times 3^6 = 3^8$. Can you see the pattern?

When we **multiply** power numbers with the **same base**, we **add the powers**.

Now look at these divisions.

$$6^7 \div 6^4 = \frac{6 \times 6 \times 6 \times \cancel{6} \times \cancel{6} \times \cancel{6} \times \cancel{6}}{\cancel{6} \times \cancel{6} \times \cancel{6} \times \cancel{6}}$$
$$= 6 \times 6 \times 6$$
$$= 6^3$$
$$4^5 \div 4^3 = \frac{4 \times 4 \times \cancel{4} \times \cancel{4} \times \cancel{4}}{\cancel{4} \times \cancel{4} \times \cancel{4}}$$
$$= 4 \times 4$$
$$= 4^2$$

So $6^7 \div 6^4 = 6^3$ and $4^5 \div 4^3 = 4^2$. Again, there is a clear pattern.

When we **divide** power numbers with the **same base**, we **subtract the powers**.

These rules work **only** when we multiply or divide numbers with the **same base**. When we multiply or divide powers with **different** base numbers, each base number must be calculated separately.

Numbers written in index form can have a **negative index**.

Look at $\frac{1}{7^3}$. We know we can write 1 as $7^0$. So $\frac{1}{7^3} = \frac{7^0}{7^3} = 7^0 \div 7^3$ and, if we use the division rules, $7^0 \div 7^3 = 7^{0-3} = 7^{-3}$. So we see that $\frac{1}{7^3} = 7^{-3}$. We follow the same rules when multiplying and dividing negative indices as we do when multiplying and dividing positive indices.

**Examples**

Calculate these. Leave your answers in index form.

a)
$$2^6 \times 2^{-3}$$
$$2^6 \times 2^{-3} = 2^{6+(-3)}$$
$$= 2^{6-3}$$
$$= 2^3$$

b)
$$3^4 \times 2^3 \times 3^{-5} \times 2^5$$
$$3^4 \times 2^3 \times 3^{-5} \times 2^5 = 3^4 \times 3^{-5} \times 2^3 \times 2^5$$
$$= 3^{4+(-5)} \times 2^{3+5}$$
$$= 3^{-1} \times 2^8$$

c)
$$10^{-4} \div 10^{-2}$$
$$10^{-4} \div 10^{-2} = 10^{-4-(-2)}$$
$$= 10^{-4+2}$$
$$= 10^{-2}$$

**Exercise 2**

1 Calculate these multiplications. Leave your answers in index form.
a) $2^5 \times 2^2$
b) $4^3 \times 4^6$
c) $6^2 \times 6$
d) $8^4 \times 8^3$
e) $9^2 \times 9^{-2}$
f) $2^{-3} \times 2$
g) $5^5 \times 5^{-7}$
h) $3^{-2} \times 3$
i) $8^{-2} \times 8^{-3}$

**2** Calculate these divisions. Leave your answers in index form.

a) $2^5 \div 2^2$

b) $4^7 \div 4^5$

c) $6^2 \div 6$

d) $8^4 \div 8^3$

e) $3^{11} \div 3^5$

f) $2^{-3} \div 2$

g) $5^5 \div 5^{-7}$

h) $11^{-2} \div 11^3$

i) $7^{-4} \div 7^{-3}$

**3** Work out these calculations. Leave your answers in index form.

a) $8^{-3} \times 8^5$

b) $7^2 \div 7^7$

c) $(2.5)^{-2} \div (2.5)^{-1}$

d) $4^3 \times 4^2 \times 4^{-5}$

e) $10^{-3} \div 10^{-2}$

f) $6^{-3} \times 6^4 \times 6^5$

g) $(0.1)^{-7} \div (0.1)^5$

h) $5^{-7} \div (5^2 \times 5^6)$

i) $4^2 \div (4^{-1} \times 4^{-2})$

**4** Work out these calculations. Leave your answers in index form.

a) $4^{-3} \times 4^5 \times 8^5 \times 8^2$

b) $4^{-1} \times 5^5 \times 5^{-7} \times 4^2$

c) $2^{-5} \times 5^3 \times 2^3 \times 5^2$

d) $3^{-1} \times 8^5 \times 3^{-2} \times 8^{-1}$

e) $10^{-3} \times 7^4 \times 10^{-2} \times 3^2 \times 3^{-5} \times 10^6 \times 7^{-2} \times 3^7$

**5** Work out these calculations. Leave your answers in index form.

a) $\dfrac{2 \times 2^5}{2^3}$

b) $\dfrac{3^5 \times 3^{-2}}{3^2}$

c) $\dfrac{2^{-5} \times 5^{-3} \times 2^3 \times 5^4}{5^{-2} \times 2^2}$

# Standard index form

Sometimes, numbers are very big or very small. Writing these numbers in the ordinary way will take a lot of time and space.

For example:

The distance from the sun to the dwarf planet Pluto is about 5 898 000 000 km.
The diameter of a blood cell is about 0.000 75 cm

To make it easier to write very big and very small numbers, we use standard index form. This is often called standard form or scientific notation.

A number written in standard form has two parts.

- The first part must be a number between 1 and 10.
- The second part is a power of 10.

These two parts are multiplied together to give the number.

- When the number is **very big**, we need to **multiply** by a power of 10, so the power (index) will be **positive**.
- When the number is **very small**, we need to **divide** by a power of 10, so the power (index) will be **negative**. (Remember: $\frac{1}{10} = 10^{-1}$)

Now we can rewrite the examples above in standard form.

For example:

The distance from the sun to the planet Pluto is about $5.898 \times 10^9$ km.
The diameter of a blood cell is about $7.5 \times 10^{-4}$ cm

It is often easier to change numbers into standard form if you write the power of 10 out in full first.

**Examples**

a) Write 370 000 in standard form.

$$370\,000 = 3.7 \times 100\,000$$
$$= 3.7 \times 10^5$$

b) Write $5.6 \times 10^7$ as an ordinary number.

$$5.6 \times 10^7 = 5.6 \times 10\,000\,000$$
$$= 56\,000\,000$$

## Multiplying and dividing numbers in standard form

When we multiply or divide numbers that are written in standard form, we do it in three steps.

**1** Multiply or divide the **number** parts. This gives the **number** part of the answer.

**2** Multiply or divide the **powers of 10**. (Remember to add the powers when you multiply and to subtract the powers when you divide.) This gives the **power of 10** part of the answer.

**3** Check that the answer is in **standard form**. Adjust the answer if you need to.

**Examples**

Calculate these. Give your answers in standard form.

a) $(8 \times 10^3) \times (4 \times 10^5)$

$$= 8 \times 4 \times 10^3 \times 10^5 \qquad \text{First change the order of the}$$
$$= (8 \times 4) \times (10^3 \times 10^5) \quad \text{calculation. We multiply the}$$
number parts and the power of
10 parts separately.
$$= 32 \times (10^3 \times 10^5) \qquad \text{Multiply the number part.}$$
$$= 32 \times 10^{3+5} \qquad\qquad \text{Multiply the power of 10 part}$$
$$= 32 \times 10^8 \qquad\qquad\quad \text{(add the powers).}$$

But 32 is not a number between 1 and 10, so this answer is **not** in standard form.

We need to rewrite the answer so that it is in standard form.

$$32 \times 10^8 = (3.2 \times 10) \times 10^8 \quad \text{Write the 32 in standard form.}$$
$$= 3.2 \times (10 \times 10^8) \quad \text{Move the brackets. Remember, we}$$
can multiply numbers in any order –
the answer will be the same.
$$= 3.2 \times 10^{8+1} \qquad\qquad \text{Multiply the powers of 10 (add the}$$
$$= 3.2 \times 10^9 \qquad\qquad\quad \text{powers).}$$

b) $(1.2 \times 10^3) \div (4 \times 10^{-8})$

$$= \frac{1.2 \times 10^3}{4 \times 10^{-8}}$$
$$= \frac{1.2}{4} \times \frac{10^3}{10^{-8}}$$
$$= (1.2 \div 4) \times (10^3 \div 10^{-8}) \qquad \text{Change the order of the calculation.}$$
$$= 0.3 \times (10^3 \div 10^{-8}) \qquad\qquad \text{Divide the number part.}$$
$$= 0.3 \times 10^{3-(-8)} \qquad\qquad\quad \text{Divide the power of 10}$$
$$= 0.3 \times 10^{11} \qquad\qquad\qquad\; \text{part (subtract the powers).}$$

But 0.3 is not a number between 1 and 10, so this answer is **not** in standard form.

$0.3 \times 10^{11} = (3.0 \times 10^{-1}) \times 10^{11}$      Write the 0.3 in standard form.

$\qquad\qquad = 3.0 \times (10^{-1} \times 10^{11})$      Move the brackets. In fact we don't need the brackets – but they help us see what is happening.

$\qquad\qquad = 3.0 \times 10^{(-1)+11}$      Multiply the powers of 10

$\qquad\qquad = 3.0 \times 10^{10}$ or $3 \times 10^{10}$      (add the powers).

## Exercise 3

1 Copy and complete this table.

| Ordinary number | Working out | Standard form |
|---|---|---|
| 300 000 | 3 × 100 000 | $3 \times 10^5$ |
| 75 000 | 7.5 × 10 000 | |
| | 8 × 100 000 000 | |
| | | $3.5 \times 10^{13}$ |
| 62 300 000 000 000 | | |
| | | $5.4 \times 10^9$ |
| | 6.93 × 10 000 000 | |
| 453 100 000 000 | | |
| | | $6.97 \times 10^5$ |
| 453 120 | | |
| | 1.097 × 100 000 | |

2 Copy and complete each equation (statement). Write a power of 10 in each space to make the equation true.

a) $0.1 = 1 \times 10^{\square}$

b) $0.000\,01 = 1 \times 10^{\square}$

c) $0.000\,000\,001 = 1 \times 10^{\square}$

d) $0.001 = 1 \times 10^{\square}$

e) $0.000\,000\,000\,01 = 1 \times 10^{\square}$

f) $0.000\,000\,000\,000\,1 = 1 \times 10^{\square}$

**3** Copy and complete this table.

| Ordinary number | Working out | Standard form |
| --- | --- | --- |
| 0.000 03 | 3 × 0.000 01 | $3 \times 10^{-5}$ |
| 0.007 5 | 7.5 × 0.001 | |
| 0.000 008 75 | 8.75 × 0.000 001 | |
| 0.000 000 003 5 | | |
| 0.000 000 000 006 23 | | |
| 0.000 000 5 | | |
| 0.000 000 047 25 | | |
| 0.05 | | |
| 0.000 007 85 | | |

**4** Write each of these numbers in standard form.
a) 300 000 000 000
b) 80 000 000
c) 700 000 000
d) 2 000 000 000
e) 42 000 000
f) 21 000 000 000
g) 3 700 000 000
h) 630
i) 3 219 000 000
j) 654 120 000
k) 897 213
l) 42 670 000 000

**5** Write each of these numbers in standard form.
a) 0.007
b) 0.000 000 002 3
c) 0.0008
d) 0.000 000 073 95
e) 0.04
f) 0.000 000 045
g) 0.0234
h) 0.000 000 002 34
i) 0.0067
j) 0.000 000 005
k) 0.3
l) 0.000 000 000 000 34

**6** Write each of these numbers as an ordinary number.

    a) $6 \times 10^5$

    b) $2 \times 10^3$

    c) $5 \times 10^7$

    d) $9 \times 10^8$

    e) $3.7 \times 10^9$

    f) $2.8 \times 10^1$

    g) $9.9 \times 10^{10}$

    h) $7.1 \times 10^4$

    i) $3.97 \times 10^2$

    j) $8.172 \times 10^2$

    k) $7.4312 \times 10^6$

    l) $1.234 \times 10^9$

**7** Write each of these numbers as an ordinary number.

    a) $3.5 \times 10^{-1}$

    b) $5 \times 10^{-4}$

    c) $7.2 \times 10^{-5}$

    d) $6.1 \times 10^{-3}$

    e) $1.17 \times 10^{-10}$

    f) $8.135 \times 10^{-7}$

    g) $6.462 \times 10^{-2}$

    h) $4.001 \times 10^{-9}$

    i) $5.5 \times 10^{-6}$

    j) $6.5 \times 10^{-8}$

    k) $3.167 \times 10^{-11}$

    l) $1.412 \times 10^{-5}$

**8** Calculate these. Give your answers in standard form.

    a) $(8 \times 10^{-5}) \times (3 \times 10^7)$

    b) $(7 \times 10^{-4}) \times (6 \times 10^{-5})$

    c) $(4 \times 10^5) \div (8 \times 10^3)$

**9** In 1992, about 1 400 000 000 steel cans and about 688 000 000 aluminium cans were recycled.
Calculate the total number of all cans that were recycled in 1992.
Give your answer in standard form.

**10** Star A is about 40 350 000 000 000 km away from our sun. Star B is even further away: it is about 15 300 000 000 000 000 km from our sun.
How much further away from the sun is Star B?
Give your answer in standard form.

11  Here are the diameters of some planets.
    Saturn $1.2 \times 10^5$ km   Jupiter $1.42 \times 10^5$ km   Pluto $2.3 \times 10^3$ km
    Write down the names of the planets in ascending order of their sizes.

12  The mass of an oxygen atom is $2.7 \times 10^{-23}$ grams. The mass of an electron is
    about 30 000 times smaller than this.
    Calculate the mass of an electron (in standard form).

# Unit 6  Other number systems

## Key vocabulary

Arabic numbers

Latin digits

lower case

mathematical operations

Roman numerals

symbol

upper case

 ## A  Our number system

The number system that we use is called the **decimal number system**. This means that every number can be made by using a combination of ten digits ('deci' = ten).

The symbols (or shapes) we use for the ten digits are often called Arabic numbers but they are more correctly called Latin digits.

# 0 1 2 3 4 5 6 7 8 9

There are many other decimal number systems. They all have ten digits, but the symbols they use look different from the Latin digits.

All decimal number systems use the same mathematical operations (+, −, ×, ÷) and all the calculations are done in the same way. It is only the symbols (or shapes) used for the digits that are different.

Sometimes you will find that some of the important bigger numbers (e.g. 100, 1000, 1 000 000 and so on) are not made out of the symbols for the ten digits. Some systems have special symbols for these bigger numbers – and sometimes even for the number 10.

## B Thai numbers

These symbols are used for the ten digits in traditional Thai writing.

| 0 | 1 | 2 | 3 | 4 | 5 | 6 | 7 | 8 | 9 |

The Thai people do not often use these traditional symbols for the ten digits. It seems that they are most often used when it is important that foreign people cannot read the numbers!

## C Chinese numbers

Here are the symbols used for the ten digits in Chinese.

| 0 | 1 | 2 | 3 | 4 | 5 | 6 | 7 | 8 | 9 |

零 一 二 三 四 五 六 七 八 九

The Chinese use special symbols for the numbers 10, 100, 1000, 10 000 and 1 000 000.

| 10 | 100 | 1000 | 10 000 | 1 000 000 |

十 百 千 万 百万

## D Japanese numbers

Here are the symbols used for the ten digits in Japanese.

れい いち に さん よん
こ ろく な はち きゅう

The Japanese use special symbols for the numbers 10, 100, 1000, 10 000 and 100 000 000.

| 10 | 100 | 1000 | 10 000 | 100 000 000 |
|----|-----|------|--------|-------------|
| じゅう | ひゃく | せん | まん | おく |

## E Ancient Egyptian numbers

The ancient Egyptians also used a decimal number system.
The numbers from 1 to 9 are made using the same number of short lines. It is not clear what symbol they used for 0.

They made other numbers by using different picture symbols.

| 10 | A yoke for cattle |
|----|-------------------|
| 100 | A coil of rope |
| 1000 | A lotus plant |
| 10 000 | A finger |
| 100 000 | A tadpole or frog |
| 1 000 000 | The figure of a god with arms raised above his head |

The rules for reading and writing ancient Egyptian numbers are quite simple.

- The higher number is always written in front of the lower number – so read from left to right.
- Where there is more than one row of picture symbols, start at the top.

Look at the following examples.

= 3244

= 21 237

# F  Ancient Greek numbers

The ancient Greeks did not use digits (0–9) as we do today.
Instead, they used their alphabet to make numbers as well as words.
For example, the **alpha** character (α) also stood for the number 1.
Here is a list of the numbers and their corresponding Greek symbols.

| 1 | 2 | 3 | 4 | 5 | 6 | 7 | 8 | 9 | 10 |
|---|---|---|---|---|---|---|---|---|---|
| $\alpha'$ | $\beta'$ | $\gamma'$ | $\delta'$ | $\varepsilon'$ | $\varsigma'$ | $\zeta'$ | $\eta'$ | $\theta'$ | $\iota'$ |

| 11 | 12 | 13 | 14 | 15 | 16 | 17 | 18 | 19 | 20 |
|---|---|---|---|---|---|---|---|---|---|
| $\iota\alpha'$ | $\iota\beta'$ | $\iota\gamma'$ | $\iota\delta'$ | $\iota\varepsilon'$ | $\iota\varsigma'$ | $\iota\zeta'$ | $\iota\eta'$ | $\iota\theta'$ | $\kappa'$ |

| 21 | 30 | 40 | 50 | 60 | 70 | 80 | 90 |
|---|---|---|---|---|---|---|---|
| $\kappa\alpha'$ | $\lambda'$ | $\mu'$ | $\nu'$ | $\xi'$ | $o'$ | $\pi'$ | $\varphi'$ |

| 100 | 200 | 300 | 400 | 500 | 600 | 700 | 800 | 900 |
|---|---|---|---|---|---|---|---|---|
| $\rho'$ | $\sigma'$ | $\tau'$ | $\upsilon'$ | $\varphi'$ | $\chi'$ | $\psi'$ | $\omega'$ | $\uparrow$ |

| 1000 | 2000 | 3000 | 10 000 | 20 000 | 100 000 |
|---|---|---|---|---|---|
| $,\alpha$ | $,\beta$ | $,\gamma$ | $,\iota$ | $,\kappa$ | $,\rho$ |

Many of these Greek symbols are still used in maths today.
The most well known is π, which is used to calculate the area and
circumference of circles. You will learn some of the other Greek
symbols when you study more advanced mathematics.

# G  Ancient Babylonian numbers

The people of ancient Babylon (who lived in the area now known as
Iraq in the Middle East) were probably the first people to develop a
written number system, in 3100BC. They recorded numbers by
pressing the differently-shaped symbols into clay tablets.
The Babylonian number system was quite simple. It used only two
symbols, both made with a wedge (or triangle). The wedge pointed
down for 1 and left for 10.

Here are some of the ancient Babylonian numbers.

| 1 | 2 | 3 | 4 | 5 | 6 | 7 | 8 | 9 | 10 |
|---|---|---|---|---|---|---|---|---|----|
| ▽ | ▽▽ | ▽▽▽ | ▽▽▽ ▽ | ▽▽▽ ▽▽ | ▽▽▽ ▽▽▽ | ▽▽▽ ▽▽▽ ▽ | ▽▽▽ ▽▽▽ ▽▽ | ▽▽▽ ▽▽▽ ▽▽▽ | ◁ |

| 20 | 30 | 40 | 50 |
|----|----|----|----|
| ◁◁ | ◁◁◁ | ◁◁ ◁ ◁ | ◁◁ ◁◁ ◁ |

The ancient Babylonians made their numbers following a 'base 10' system. This means that they used tens and unit, just as we do.

So ◁ ▽ ▽ ▽ = 10 + 3 = 13

and ◁◁ ▽ ▽ ▽ / ◁ ▽ ▽ ▽ / ◁ ▽ = 40 + 7 = 47

However, their number system was actually 'base 60', which means that they only used these symbols to make the numbers 1 to 59 (plus they used a space for zero). To make larger numbers, they used a different system that we are not going to learn about here.

The wedge or triangle symbols for the numbers were often 'joined up' in their groups so they would probably have looked more like this.

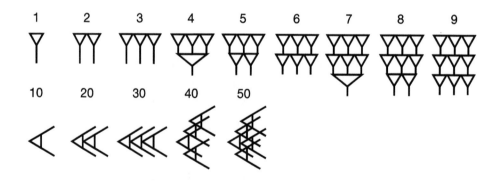

The Babylonians were also one of the first peoples to have a calendar. Their days were divided into 24 hours, each hour was divided into 60 minutes, and each minute was divided into 60 seconds. This form of counting has survived for 4000 years and is still used today to measure time all over the world.

#  Roman numbers

About 2000 years ago, the Romans ruled large parts of what we now know as Europe. Their culture and civilisation was very advanced. The system they used to write their numbers is still often used all over the world today as a 'second' number system. The Roman numbers are known as Roman numerals.

The Romans used some of the letters of their alphabet to write their numbers (like the Greeks). They used the letters with simple shapes so that it was easy to carve them into stone! They used the letters in both lower case and upper case.

Here are some of the letters they chose for their numbers:

|  | 1 | 5 | 10 | 50 | 100 | 500 | 1000 |
|---|---|---|---|---|---|---|---|
| upper case | I | V | X | L | C | D | M |
| lower case | i | v | x | l | c | d | m |

To make their numbers, the Romans followed some simple rules.
- You can only use a maximum of **three** of the same letter together.
- If there is a smaller number **before** a bigger number, you **take away** the smaller one from the bigger one.

The number four shows us how these rules go together: the Romans did not write the number four as IIII; instead they used a I **before** the V to show that 1 is taken away from the 5! So the number 4 is written as IV.

We must remember that we take away the smaller number only if it is **before** the bigger number. If the smaller number is **after** the bigger number, we add. For example:

XC stands for C − X = 100 − 10      The smaller number (X = 10) is
                   = 90            before the bigger number
                                    (C = 100).

CX stands for C + X = 100 + 10      The smaller number is after the
                   = 110        bigger number.

The numbers from 1 to 10 look like this.

|  | 1 | 2 | 3 | 4 | 5 | 6 | 7 | 8 | 9 | 10 |
|---|---|---|---|---|---|---|---|---|---|---|
| upper case | I | II | III | IV | V | VI | VII | VIII | IX | X |
| lower case | i | ii | iii | iv | v | vi | vii | viii | ix | x |

**Exercise 1**

1 Write each of these Roman numerals in modern decimal numbers.
   a) XII
   b) XXIX
   c) xli
   d) LXXVII
   e) XCV
   f) civ
   g) cccxl
   h) CDXVI
   i) dcclxviii
   j) MDCLXXXIV
   k) mmcdxxxvi

2 Write each of these numbers using Roman numerals (use upper case).
   a) 25
   b) 52
   c) 68
   d) 99
   e) 146
   f) 555
   g) 941
   h) 1033
   i) 1939
   j) 1992
   k) 2537

3 Here is a shopping list written with Roman numerals. Copy the list using modern numbers.
   a) Pack of XII toilet rolls
   b) VI apples
   c) Breakfast cereal (D grams)
   d) Orange squash (DCCL millilitres)
   e) CCCXX grams of cheese
   f) Roll of XXIV black rubbish bags
   g) VIII slices of bacon
   h) Pack of V CLXXX-minute videotapes
   i) MD gram tub of ice cream
   j) CCXL tea bags

**Project activity**

Imagine that you are an alien from another planet. On your planet, you use these symbols for your numbers: I = 1, T = 2, S = 20, B = 200.

So counting on your planet would start like this: I, T, TI, TT, ...

RULE: You cannot write more than **five** of the same symbols together.

Write out the rest of your planet's numbers, up to 200.

# Unit 7 Approximation and estimation

## A Approximation

When we solve a maths problem, it is important that our work is accurate and precise; and also that our answer is exact and correct.

In real-life situations, it is often not so important to know the exact answer to a calculation. Sometimes, it is more important to know **roughly** what the answer is (so that we know how much cash to take with us, for example; or how much food to make for the party, and so on). The answer that we use is **not exactly** the correct answer, but **nearly** the correct answer – one that is **close** but not exact. Answers that are nearly correct are called approximate answers.

Just **how close** the approximate answer is to the precise answer depends on **how accurate** we need to be. We can use a simpler number if we don't need to give all the details.

One of the ways we can make an approximate answer is by rounding the numbers we use in the calculation. Whether we round or not depends on the 'story' behind the calculation – what we are working out, and why. For example:

21 152 people attend a football match.

- A newspaper report might say '21 000 people watched Real Betis beat Real Murcia 2:0 in last night's match'. The newspaper's readers don't need to know exactly how many people were at the match – the rounded number of 21 000 (rounded to the nearest 1000) is close enough, and is easier to remember.

● The accountant who checks the money earned from selling the tickets to watch this football match needs to know exactly how many people were there – an approximate figure is not good enough.

## Rounding to the nearest 10, 100, 1000, ...

Look at the number 7487.

This number can be rounded off to different degrees of accuracy.

### Rounding to the nearest 10

7487 is between 7480 and 7490, but it is **closer** to 7490.
7487 rounded to the **nearest 10** is 7490.

### Rounding to the nearest 100

7487 is between 7400 and 7500, but it is **closer** to 7500.
7487 rounded to the **nearest 100** is 7500.

### Rounding to the nearest 1000

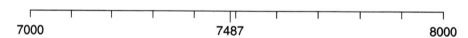

7487 is between 7000 and 8000, but it is **closer** to 7000.
7487 rounded to the **nearest 1000** is 7000.

Depending on the 'story' and the **degree of accuracy** we need, the number 7487 can be rounded to 7490, 7500 or 7000.

NOTE: If the number is exactly in the middle between two rounding numbers, we round the number up. For example:

75 rounded to the nearest 10 is 80.
450 rounded to the nearest 100 is 500.
8500 rounded to the nearest 1000 is 9000.

## Rounding to a power of 10

We can round numbers to any power of 10 without using a number line simply by looking at the first unwanted digit.

- If the digit is 5 or more, round **up**: increase the last digit you need by 1 and replace the following digit(s) by 0.
- If the digit is less than 5, round **down**: leave the last digit you need as it is and replace the following digit(s) by 0.

**Examples**

Round the number 6582

a) to the nearest 10

We are rounding to the nearest 10, so we look at the digit in the units column: **2**

This is less than 5, so round **down**.

6582 to the nearest 10 is 6580.

b) to the nearest 100

We are rounding to the nearest 100, so we look at the digit in the tens column: **8**

This is more than 5, so round **up**.

6582 to the nearest 100 is 6600.

c) to the nearest 1000

We are rounding to the nearest 1000, so we look at the digit in the hundreds column: **5**

We round 5 **up**.

6582 to the nearest 1000 is 7000.

If we are given a rounded figure, we cannot work out what the exact figure was, but we can work out the smallest and biggest numbers it could be.

**Examples**

The exact number of pupils at Tesaban 6 School in Chiang Rai has been rounded to the nearest 100 to give a number of 2600 pupils.

a) What is the **smallest possible** number of pupils at Tesaban 6 School?

b) What is the **biggest possible** number of pupils at Tesaban 6 School?

Which numbers would be rounded to 2600 as the nearest 100?

2549 would be rounded down to 2500

2550 would be rounded up to 2600.

Every number after 2550, up to 2649, will be rounded up to 2600.

2650 would be rounded to 2700.

a) The smallest possible number of pupils is 2550.

b) The biggest possible number of pupils is 2649.

**Exercise 1**

**1** a) Which of these numbers is closest to 4872?
4850, 4860, 4870, 4880, or 4890

b) Which of these numbers is closest to 4872?
4600, 4700, 4800, 4900, or 5000

c) Which of these numbers is closest to 4872?
3000, 4000, 5000, or 6000

**2** Round the number 7425
a) to the nearest 10
b) to the nearest 100
c) to the nearest 1000.

**3** Round each of these numbers
i) to the nearest 10
ii) to the nearest 100
iii) to the nearest 1000.
a) 7613       b) 977
c) 61 115       d) 9714
e) 623       f) 9949
g) 5762       h) 7501
i) 7500       j) 7499

**4** Round each of these numbers. Decide from the story how accurate it needs to be.
a) There were 19 141 people at the football match.
b) There were 259 people on the plane.
c) Tom has 141 marbles.
d) The class raised $49.67 for charity.
e) There are 229 pupils in Year 7.
f) The population of Bangkok is 5 572 955.
g) The land area of Thailand is 513 123 km².
h) The distance from Bangkok to London is 10 665 km.
i) Sarah spent €50.99 on holiday souvenirs.
j) There are 693 pupils in the school.

5 Write down one example of a number that will fit each of these roundings.
   a) The number is 750 when rounded to the nearest 10, but it is 700 when rounded to the nearest 100.
   b) The number is 750 when rounded to the nearest 10 but it is 800 when rounded to the nearest 100.
   c) The number is 8500 when rounded to the nearest 100 but it is 8000 when rounded to the nearest 1000.
   d) The number is 8500 when rounded to the nearest 100 but it is 9000 when rounded to the nearest 1000.

6 The number in this newspaper report was correct to the nearest 1000: '43 000 spectators watched the thrilling cricket Test Match.' What is the **smallest** possible number of spectators?

7 Carl has 140 postcards in his collection. This number is correct to the nearest 10.
   a) What is the smallest number of postcards Carl could have in his collection?
   b) What is the biggest number of postcards Carl could have in his collection?

8 'You need 2700 tiles to tile your roof.' This number is correct to the nearest 100. What is the biggest number of tiles that might be needed?

## Rounding using decimal places

If material costs $1.99 a metre, how much will 1.75 metres cost?

We find the answer by calculating $1.75 \times 1.99$. The answer will have 4 decimal places (see Unit 4).

$1.75 \times 1.99 = 3.4825$

So the material costs $3.4825 or 348.25 cents.

This is really a silly answer! There is no way to pay 0.25 cents (the smallest coin is 1¢). We would actually pay $3.48, and the other decimal places are meaningless.

On other occasions, a different number of decimal places might be more appropriate. As before, we have to decide what to do by thinking about the story behind the calculation. This is called choosing a **sensible** degree of accuracy.

## Rounding to a number of decimal places

Sometimes you will be told how many decimal places the answer to a question should have; sometimes you will have to decide for yourself. Follow these rules to round your answer.

1 Write the number you have calculated with **one more** decimal place than you need.
2 Look at the digit in the **last** decimal place.
   ● If the digit is 5 or more, round **up**: increase the last digit you need by 1 and 'drop' the digit in the last decimal place.
   ● If the digit is less than 5, round **down**: leave the last digit you need as it is and 'drop' the digit in the last decimal place.
3 Remember to include the correct **units** in your answer, and state **how accurate** it is (e.g. 'correct to 2 decimal places' or 'correct to 2 d.p.').

**Examples**

Round each of these numbers to the number of decimal places given in brackets.

a) 2.764 35 (2 d.p.)
   Write the number with 1 more decimal place than you need: 2.764
   Look at the digit in the last decimal place: 4
   This is less than 5, so round **down**.
   2.764 35 correct to 2 d.p. is 2.76.

b) 2.762 85 (3 d.p.)
   Write the number with 1 more decimal place than you need: 2.7628
   Look at the digit in the last decimal place: 8
   This is more than 5, so round **up**.
   2.762 85 correct to 3 d.p. is 2.763.

A rounded number should always have the number of decimal places stated. This means that if we round a number to, for example, 2 decimal places, we must write 2 decimal places – even if this means writing one or more zeros (0) in the last decimal places that we would not normally write.

**Examples**

Round each of these numbers to the number of decimal places given in brackets.

a) 7.104 (2 d.p.)
   Write the number with 1 more decimal place than you need: 7.104
   Look at the digit in the last decimal place: 4
   This is less than 5, so round **down**.
   1.704 correct to 2 d.p. is 1.70. (We must write the zero at the end.)

b) 5.98 (1 d.p.)
   Write the number with 1 more decimal place than you need: 5.98
   Look at the digit in the last decimal place: 8
   This is more than 5, so round **up**.
   5.98 correct to 1.d.p. is 6.0. (We must write the zero at the end.)

**Exercise 2**

**1** Write the number 3.9617 correct to
a) 3 decimal places
b) 2 decimal places
c) 1 decimal place.

**2** Write the number 567.654 correct to
a) 2 decimal places
b) 1 decimal place
c) the nearest whole number.

**3** Round each of these numbers to the number of decimal places given.
a) 2.367 to 1 decimal place
b) 0.964 to 2 decimal places
c) 0.965 to 2 decimal places
d) 15.2806 to 3 decimal places
e) 0.056 to 2 decimal places
f) 4.991 to 2 decimal places
g) 4.996 to 2 decimal places
h) 17.55493 to 2 decimal places
i) 17.55493 to 3 decimal places

**4** Write each of these decimals correct to 2 decimal places.
a) 4.834
b) 1.641
c) 6.978
d) 2.887
e) 14.055
f) 28.065
g) 4.8319
h) 2.8427
i) 7.0842
j) 17.0413
k) 0.98932
l) 0.27642

**5** Round each of these decimals to the number of decimal places given in brackets.
a) 4.87 (1 d.p.)
b) 12.843 (2 d.p.)
c) 0.0475 (3 d.p.)
d) 6.79 (1 d.p.)
e) 0.9408 (3 d.p.)
f) 0.8736 (3 d.p.)
g) 18.687 (2 d.p.)
h) 4.0649 (3 d.p.)
i) 2.94 (1 d.p.)
j) 17.63 (1 d.p.)
k) 3.863 (1 d.p.)
l) 9.877 (1 d.p.)
m) 24.938 (1 d.p.)
n) 0.145 (1 d.p.)
o) 7.9327 (2 d.p.)
p) 73.0645 (3 d.p.)
q) 0.00762 (3 d.p.)
r) 2.0419 (2 d.p.)
s) 14.83333 (2 d.p.)
t) 4.9128 (2 d.p.)

**6** Calculate each of these divisions. Give your answers correct to 3 decimal places.
a) $0.47 \div 0.3$
b) $0.83 \div 0.6$
c) $0.068 \div 0.07$
d) $0.082 \div 0.03$
e) $0.53 \div 0.006$
f) $0.29 \div 0.007$
g) $3.61 \div 1.1$
h) $7.58 \div 1.2$
i) $6.56 \div 0.09$

**7** For each of these calculations,
  i) work out the correct answer, up to 4 decimal places
  ii) write the answer correct to 3 decimal places
  iii) write the answer correct to 2 decimal places
  iv) write the answer correct to 1 decimal place.

a) 6.12 × 7.54
b) 89.1 × 0.67
c) 90.53 × 6.29
d) 98.6 ÷ 5.78
e) 6.72 ÷ 10.14
f) 1.6074 ÷ 0.47

**8** Work out these calculations. Write each answer using a sensible degree of accuracy.
a) 1 metre of cloth costs $6.99. How much does 1.74 metres cost?
b) 1 kg of cheese costs $5.21. How much does 0.454 kg cost?
c) A pack of 6 CDs costs $7.99. How much does one CD cost?
d) Petrol costs 98.4 cents a litre. How much will I pay for 15.6 litres?

## Rounding using significant figures

Look at this number: 0.000 364 907

What happens if we try to make it simpler by **rounding** it using the methods we have learned on page 132?

Correct to the nearest whole number, the answer is 0
Correct to 3 decimal places, the answer is 0.000

These are not very useful numbers!

Instead we must round to a number of **significant figures**. 'Significant' means 'important', and 'figures' are also called 'digits', so we are talking about the **most important digits**.

In any number, the most important digit is the one with the biggest place value. For example, in the number 237, the digit 2 has the biggest place value. It is worth 200. So 2 is the most **significant figure**. We also call it the first significant figure. 3 is the second significant figure and 7 is the third significant figure.

In a number less than zero, the most important digit is the first digit that is not zero. The zero digits before it are important, because they act as **place holders**, so that each digit is in the correct place. However, they are not **significant** because their **place value** is **0**. For example, in the number 0.003 28, the digit 3 has the biggest place value. It is worth $\frac{3}{1000}$. So 3 is the most significant figure. The zeros before the 3 are important because they keep the 3 in the thousandths column, but they are not significant.

## Rounding to a number of significant figures

We can use these rules to round to any number of significant figures we choose.

1 Find the most significant figure in the number.
2 Count the number of significant figures you need.
3 Look at the next digit after the last significant figure you need.
   - If the digit is 5 or more, round **up**: increase the last significant figure by 1.
   - If the digit is less than 5, round **down**: leave the last significant figure as it is.
4 Make sure that the decimal point stays in the same place and that all the digits keep the same place value.
   - If the number is greater than 1, look at the digits before the decimal place. Fill all the places between the last significant figure and the decimal points with zeros. Any digits after the decimal point can just be ignored.
   - If the number is less than 1, make sure that you write the same number of zeros before the first significant figure as there were in the unrounded number. Any digits after the last significant figure can just be ignored.
5 Remember to include the correct **units** in your answer, and state **how accurate** it is (e.g. 'correct to 2 significant figures' or 'correct to 2 s.f.').

**Examples**

Round each of these numbers to the number of significant figures given in brackets.

a) 4 500 732.019 4 (2 s.f.)

The most significant figure is the first 4. It has a place value of 4 000 000.
We want 2 significant figures: 4 500 732.019 4
Look at the next digit: **0**
This is less than 5, so round **down**. The second significant figure, **5**, remains the same.
The number is more than 1, so we must fill all the places between the 5 and the decimal places with zeros. We can check we have the right number of zeros by making sure that the 4 still has a place value of 4 000 000.
4 500 732.019 4 correct to 2 significant figures is 4 500 000.

b) 0.000 364 907 (1 s.f.)

The most significant figure is the 3. It has a place value of $\frac{3}{10\,000}$.
We want only this 1 significant figure: 0.000 364 907
Look at the next digit: **6**
This is more than 5, so round **up**. The significant figure becomes 4.
The number is less than 1, so we must make sure we write the same number of zeros before the first significant figure as there were in the unrounded number. We can check we have the right number of zeros by making sure the 4 has a place value of $\frac{4}{10\,000}$.
0.000 364 907 correct to 1 significant figure is 0.0004.

### Significant zeros

We already know that, in a number less than 1, the zeros on the left are **not** significant.

When we are rounding a number, we might also find zeros **after** the first significant figure.

● Zeros between non-zero digits. These zeros **are** significant.

For example:

20.013 correct to 4 significant figures is 20.01. The two zeros are the second and third significant figures.
11.042 correct to 3 significant figures is 11.0. The zero is the third significant figure.

● Zeros at the end of the number. These zeros may be significant, depending on the degree of accuracy you are asked to use.

For example:

1.000 correct to 3 significant figures is 1.00. Look at the unrounded number: the first two zeros **are** significant; the third zero is **not** significant.
23 000 correct to 3 significant figures is 23 000. The first zero **is** significant; the last two zeros are **not** significant (they are just place holders).

### Choosing a sensible number of significant figures

When we do calculations that use measurements, the answer should not be given to a greater degree of accuracy than the measurements used in the calculation. For example, if the measurements are correct to 2 significant figures, the answer must not have more than 2 significant figures. Usually, we would give the answer to the **same degree of accuracy** as the measurements we have used.

**Example**

A rectangle is 4.6 cm wide and 7.2 cm long. The measurements are correct to 2 s.f. To calculate the area of the rectangle, we multiply the length by the width.

$$4.6 \times 7.2 = 33.12$$
$$= 33 \text{ (correct to 2 s.f.)}$$

The area of the rectangle is 33 cm², correct to 2 s.f.

### Exercise 3

**1** Write each of these numbers correct to 1 significant figure.
  a) 17                    b) 523
  c) 0.34                  d) 0.019
  e) 24.6                  f) 30.08

**2** Write each of these numbers correct to 3 significant figures.
  a) 2732                  b) 3059
  c) 0.012 43              d) 0.031 58
  e) 42 617                f) 86 279
  g) 239 821               h) 1 097 288
  i) 0.007 008 3           j) 0.000 496 81

**3** Write each number
  i) correct to 2 decimal places
  ii) correct to 2 significant figures.
  a) 1.8181
  b) 0.0708
  c) 8.057
  d) 4567.123

**4** Round each of these numbers to the number of significant figures written in brackets.
  a) 456 000 (2 s.f.)       b) 454 000 (2 s.f.)
  c) 7 981 234 (3 s.f.)     d) 7 981 234 (2 s.f.)
  e) 1290 (2 s.f.)          f) 19 602 (1 s.f.)
  g) 0.000 567 (2 s.f.)     h) 0.093 748 (2 s.f.)
  i) 0.093 748 (3 s.f.)     j) 0.093 748 (4 s.f.)
  k) 0.010 245 (2 s.f.)     l) 0.029 94 (2 s.f.)

**5** A field is 18.6 m wide and 25.4 m long. These measurements are correct to 3 significant figures. Calculate the area of the field by multiplying the length by the width. Write your answer to a sensible degree of accuracy.

6 For each of these calculations,
   i) work out the correct answer
   ii) write the answer correct to 4 significant figures
   iii) write the answer correct to 3 significant figures
   iv) write the answer correct to 2 significant figures.
   a) $672 \times 123$
   b) $6.72 \times 12.3$
   c) $78.2 \times 12.8$
   d) $7.19 \div 987.5$
   e) $124 \div 65\,300$
   f) $1.6074 \div 0.47$

## Rounding and real-life problems

In a real-life problem, numbers must be rounded to give the most **sensible** answer for the story. It is important to use your common sense when you think about these problems!

**Examples**

a) Penny is arranging a barbeque party for her friends. There will be 50 people in total. She wants to buy enough food so that there is **one burger** for every person.
The supermarket sells burgers in packs containing 12 burgers. How many packs of burgers must she buy?

Number of packs needed $= 50 \div 12$
$\qquad\qquad\qquad\qquad = 4.1\dot{6}$

But Penny cannot buy $0.1\dot{6}$ of a pack (we can only buy **whole** packs). So we need to round the answer to a **whole number** of packs.

Using the rule we learned above, $4.1\dot{6}$ correct to the nearest whole number is 4. But think about the story: if Penny buys 4 packs of burgers she will have $4 \times 12 = 48$ burgers; two of her guests will not get a burger. It would actually be most sensible to round **up** to 5, even though that is **not** the **nearest** whole number.
If Penny buys 5 packs of burgers, she will have $5 \times 12 = 60$ burgers; each of her 50 friends will get a burger, and 10 hungry guests can have a second burger.

Penny must buy 5 packs of burgers.

b) The Year 7 pupils at a school in Germany are going on a school trip to Berlin Zoo. The total number of pupils and teachers going on the trip is 242. Each bus can carry 55 passengers. How many buses will the school need?

$242 \div 55 = 4.4$ or 4 remainder 22

If we round **down** to the nearest whole number, the school will get 4 buses and 22 people will have to be left behind.
We must round **up** instead: if the school gets 5 buses, all the pupils and teachers can go, and there will be 33 spare seats.

The school will need 5 buses.

c) Tai Po School has a new computer room. The computer desks are to be placed around the walls. The available wall space is 28.64 m. Each computer desk is 80 cm wide. How many desks will fit into the room?

$28.64 \, m = 2864 \, cm$
$2864 \div 80 = 35.8$

Even though the answer is nearly 36 desks, the 36th desk would not fit in, and we cannot have 0.8 of a desk!

35 desks will fit in the room.

 **Exercise 4**

1 49 pupils are waiting to go to the sports stadium. A minibus can take 15 passengers. How many trips must the minibus make?

2 A classroom wall is 700 cm long. A table is 120 cm long. How many tables will fit along the wall?

3 76 people are waiting to get to the top of the Eiffel Tower in Paris. The lift can take only 8 people at a time. How many times must the lift go up?

4 A group of 175 people are going to Mae Sai. A bus can take 39 passengers. How many buses will be needed?

5 There are 210 pupils in Year 7. Exercise books are sold in packs of 25. How many packs will be needed so that every pupil gets 2 exercise books?

6 The car park at the supermarket is 62 m wide. A parking space must be 2.5 m wide. How many parking spaces will fit in?

7 A factory puts 17 sweets into a bag. The factory has 520 sweets. How many bags can the factory fill?

8 It costs 32p to post a letter in London. I have £5. How many letters can I post?

9 Oranges cost 27¢ each. I have $1.50. How many oranges can I buy?

10 Kim needs 25 candles for her birthday cake. One pack contains 4 candles. How many packs must she buy?

11 Lauren wants to buy 28 doughnuts for a party. Doughnuts are sold in packs of 12. How many packs must she buy?

## B Estimation

So far in this unit we have learned how to **round** the exact answer to a calculation so that we have an **approximate** answer that is simpler to use.

Now we are going to learn how to find a simple answer to a calculation **before** we know what the exact answer is.

If we **guess** what the answer is, our guess may be correct, or it may be close to the exact answer, or it may be completely wrong! Guessing is not very reliable and we cannot be sure our guess is even close until we work out the exact answer.

Instead, we can round each of the numbers in the calculation. This makes the calculation much easier and quicker, and we can be sure that the answer will be **close** to the exact answer. We call this simplified calculation an **estimate**.

We can summarise these two processes like this.

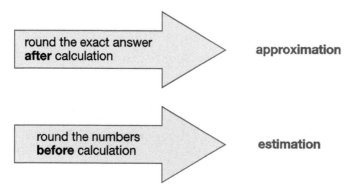

round the exact answer **after** calculation → approximation

round the numbers **before** calculation → estimation

Estimation is very useful. For example, if we are going shopping, we can make sure we take enough cash with us by quickly estimating what we will spend. We can also use estimates to check that we get the right amount of change.

An estimate is also used to check whether the exact answer to a calculation is about the right 'size'. If the exact answer is very different from the estimate, then you have probably made a mistake!

The rules for making an estimation of a calculation are very simple.
1 Round off every number in the calculation to 1 significant figure.
2 Do the calculation with these simpler, rounded numbers.

**Examples**

For each of these calculations,
i) estimate the answer
ii) work out the exact answer (it should be close to your estimate).

a) $421 \times 48$
   i) Round 421 to 1 significant figure: 400
      Round 48 to 1 significant figure: 50
      $400 \times 50 = 20\,000$
   ii) $421 \times 48 = 20\,208$

b) $608 \div 19$
   i) Round 608 to 1 significant figure: 600
      Round 19 to 1 significant figure: 20
      $600 \div 20 = 30$
   ii) $608 \div 19 = 32$

c) $\dfrac{78.5 \times 0.51}{18.7}$
   i) Round 78.5 to 1 significant figure: 80
      Round 0.51 to 1 significant figure: 0.5
      Round 18.7 to 1 significant figure: 20
      $$\frac{80 \times 0.5}{20} = \frac{40}{20}$$
      $$= 2$$
   ii) $\dfrac{78.5 \times 0.51}{18.7} = 2.14\dot{0}\dot{9}$

**Exercise 5**

**1** For each of these calculations,
   i) estimate the answer
   ii) work out the exact answer using long multiplication.
   a) 32 × 41
   b) 12 × 66
   c) 58 × 34
   d) 72 × 45
   e) 34 × 78
   f) 17 × 219
   g) 291 × 56
   h) 312 × 23

**2** For each of these calculations,
   i) estimate the answer
   ii) work out the exact answer using long division.
   a) 594 ÷ 18
   b) 609 ÷ 21
   c) 256 ÷ 16
   d) 840 ÷ 35

**3** For each of these calculations,
   i) estimate the answer
   ii) work out the exact answer using long multiplication.
   a) 4.2 × 1.8
   b) 8.9 × 3.1
   c) 48.1 × 4.2
   d) 103.4 × 2.9

**4** For each of these calculations,
   i) estimate the answer
   ii) work out the exact answer using long division.
   a) 10.78 ÷ 4.9
   b) 19.68 ÷ 4.1
   c) 30.4 ÷ 3.2
   d) 203.49 ÷ 5.1

**5** Kafil wants to build a conservatory costing $5328. The new furniture for the conservatory will cost $784.
   a) Estimate the total amount that Kafil plans to spend.
   b) Work out the exact total amount that Kafil will spend.

**6** During the first hour after a shop opened, 463 people walked inside.
During the second hour, another 1273 people walked into the shop.
   a) Estimate how many people walked into the shop during the first 2 hours.
   b) The owner would like 3000 people to visit his store in the first 3 hours.
      Estimate how many more people must enter during the 3rd hour for this
      to be true.

**7** For each of these calculations,
   i) estimate the answer
   ii) work out the exact answer using a calculator.
   a) $\dfrac{7.9 \times 3.9}{4.8}$
   b) $\dfrac{403 \times 0.29}{6.2}$
   c) $\dfrac{81.7 \times 4.9}{1.9 \times 10.3}$
   d) $\dfrac{98.7 \times 32.1}{64.5 \times 4.85}$

# Unit 8 Measures and measurement

## A Systems of measurement

It is important to know 'how much' of something we have, or 'how many' things there are.

We already know the **counting numbers**, which we use to tell us how many things there are.

However, we cannot use the counting numbers to tell us the **size** of something.

Think about these questions.
- How do we tell **how long** a piece of wood is?    length
- How do we tell **how heavy** a box of books is?    mass
- How do we tell **how long** a football match lasts?    time

● How do we tell **how much** water is in the bottle?    volume or capacity

● How do we tell **how fast** a motor car is going?    speed

Length, mass, time, volume (or capacity) and speed are all measures; we use measures to tell us how much we have of different things. People have used many different ways to measure things – for example, the length of a thumb, the length of a foot and the length of a pace (walking step) have all been used to measure the length of things.

These measures can be confusing, however. I might tell you that my office is 34 of 'my feet' long, or that my desk is 65 of 'my thumbs' wide. This is fine for me, but how will you or anybody else know how big my foot or my thumb is?

This is why everyone has agreed to use the same references or units for different types of measurement. There are two common systems of measurement.

● The Metric system is used all over the world. It is a **decimal** system – everything is based on multiples of 10 and powers or 10.

● The Imperial system is an older system that is still widely used in England and some other countries. This is not a decimal system, and so it is much more difficult to use. We will learn more about the Imperial system in Level 2.

## B Time

There is one measurement that we use every day but is **not** based on a decimal system: time. The system of units for measuring time was developed by the Babylonian people many hundreds of years ago. It is based on a unit of 60. There is no real pattern to the system, but most of us learn it very young!

### Units of time

| | |
|---|---|
| 60 seconds (s) | = 1 minute (min) |
| 60 minutes | = 1 hour (h) |
| 24 hours | = 1 day |
| 7 days | = 1 week |
| 28–31 days | = 1 month |
| 365 days | = 1 year (approx.) |
| 52 weeks | = 1 year (approx.) |
| 12 months | = 1 year |
| 366 days | = 1 leap year |
| 10 years | = 1 decade |
| 100 years | = 1 century |

Use this rhyme to help you remember the number of days in each month.

30 days has September, April, June and November. All the rest have 31, except for February alone, which has 28 days clear and 29 in each leap year.

**Examples**

**a)** Find the number of minutes in

    **i)** $\frac{1}{3}$ hour             **ii)** $3\frac{3}{4}$ hours.

    **i)** 1 hour = 60 minutes

$$\frac{1}{3} \text{ hour} = \frac{1}{3} \times 60 \text{ minutes}$$
$$= 20 \text{ minutes}$$

    **ii)** 1 hour = 60 minutes

$$3\frac{3}{4} \text{ hours} = 3\frac{3}{4} \times 60 \text{ minutes}$$
$$= \frac{15}{4} \times 60 \text{ minutes}$$
$$= 225 \text{ minutes}$$

**b)** Find the number of hours in

    **i)** $2\frac{3}{8}$ days

    **ii)** 2 weeks and 3 days

    **i)** 1 day = 24 hours

$$2\frac{3}{8} \text{ days} = 2\frac{3}{8} \times 24 \text{ hours}$$
$$= \frac{19}{8} \times 24 \text{ hours}$$
$$= 57 \text{ hours}$$

    **ii)** 1 day = 24 hours and 1 week = 7 days

$$2 \text{ weeks and } 3 \text{ days} = (2 \times 7) + 3 \text{ days}$$
$$= 14 + 3 \text{ days}$$
$$= 17 \text{ days}$$
$$= 17 \times 24 \text{ hours}$$
$$= 408 \text{ hours}$$

**c)** Find the number of hours in

    **i)** 150 minutes             **ii)** 195 minutes.

    **i)** 60 minutes = 1 hour

$$150 \text{ minutes} = 150 \div 60 \text{ hours}$$
$$= \frac{150}{60}$$
$$= 2\frac{1}{2} \text{ hours}$$

    **ii)** 60 minutes = 1 hour

$$195 \text{ minutes} = 195 \div 60$$
$$= \frac{195}{60}$$
$$= 3\frac{1}{4} \text{ hours}$$

Look at the clock on the right. It is divided into **12 equal sections**. Each of these sections stands for **1 hour** of time. The time is read as 1 o'clock, 2 o'clock etc. 'o'clock' is short for 'of the clock'.

Because each day has **24 hours**, not 12, we have to use the same numbers twice, once during the first part of the day and once during the second part. Somehow, we have to show the difference between them.

- Each new day starts at midnight (12 o'clock at **night**).
- The first half of the day is the part that is **before noon**. In Latin, 'before noon' is 'ante-meridian' and this is shortened to '**a.m.**'
- The second half of the day is the part that is **after noon**. In Latin, 'after noon' is 'post-meridian' and this is shortened to '**p.m.**'

NOTE: 12 o'clock in the **day** is called 'midday' or '12 noon' (sometimes shortened to just 'noon').

Look at the following examples:

Nine o'clock in the morning is **9 hours after midnight**, which is **before noon**. It is written as **9.00 a.m.**
Nine o'clock in the evening is **9 hours after noon**. It is written as **9.00 p.m.**

There is another way to talk about the time of day: the **24-hour clock**. This is used in many countries around the world. The military also prefer the 24-hour clock: they do not use the 'a.m.' and 'p.m.' in case someone does not hear correctly, and makes a mistake!

In the 24-hour clock, the day starts at midnight as usual, and each hour is counted off as before until 12 noon. The next hour is called 13, then 14, and so on. This means that there is no need to write 'a.m.' or 'p.m.' – each number can mean only one time of day or night.

Look at the following examples:

Nine o'clock in the morning is **9 hours after midnight**. It is written as 09 00 (read as 'zero nine hundred hours').
Nine o'clock in the evening is 9 + 12 = **21 hours after midnight**. It is written as 21 00 (read as 'twenty-one hundred hours').

Now that digital clocks and watches are common, the 24-hour clock is used much more widely than it used to be, even in countries that have traditionally used the 12-hour clock.

**Examples**

a) Write each of these times using the 24-hour clock.
   i) 8.25 a.m.                    ii)     7.30 p.m.

   i) 8.25 a.m. is before noon.
      In the 24-hour clock, 8.25 a.m. is 08 25.
   ii) 7.30 p.m. is after noon.      7 + 12 = 19
      In the 24-hour clock, 7.30 p.m. is 19 30.

**b)** Rewrite each of these times using a.m. or p.m.
   **i)** 05 40
   **ii)** 17 15
   **i)** 05 40 is before noon.
      So 05 40 is 5.40 a.m.
   **ii)** 17 15 is after noon (17 > 12).
      17 − 12 = 5
      So 17 15 is 5.15 p.m.

Sometimes we need to solve problems involving time. Unless you are told otherwise, you should give your answer in the same format as the question (e.g. if the times in the question are written in the 24-hour clock, you should write your answer in the 24-hour clock).

**Example**

A train leaves New Delhi at 13 00. It takes $6\frac{1}{2}$ hours to travel to Haridwar. Calculate at what time the train arrives in Haridwar.

13 00 hours + 6 30 hours = 19 30.
The train arrives in Haridwar at 19 30.

When we add times together, it is often best to add the minutes and hours separately.

**Example**

Add these two times together: 2 h 17 min and 7 h 53 min

Add the minutes first.
17 + 53 = 70 minutes
         = 1 hour + 10 minutes
Now add **all** the hours together.
2 + 7 + 1 (from the minutes) = 10 hours
So the total is 10 h 10 min.

Always remember that there are 60 seconds in a minute and 60 minutes in an hour (not 100).

**Examples**

**a)** How much time is there between 06 55 and 13 23 on the same day?

Length of time = 13 h 23 min − 6 h 55 min
Subtract the hours and the minutes separately.
We cannot take 55 minutes from 23 minutes – so borrow
1 hour = 60 minutes.
Length of time = (12 − 6) h (83 − 55) min
             = 6 h 28 min

**b)** How much time is there between 22 46 today and 09 14 tomorrow?

Today: 22 46 to midnight = 24 00 − 22 46 = 1 h 14 min
Tomorrow: midnight to 09 14 = 9 h 14 min
Total length of time = 1 h 14 min + 9 h 14 min
= 10 h 28 min

**c)** How much time is there between 4.46 p.m. and 8.13 a.m. (on the next day)?

Always change the times into 24-hour clock times before you start to calculate.
4.46 p.m. = (4 + 12) 46
= 16 46
8.13 a.m. = 08 13
From 16 46 until midnight = 24 00 − 16 46 = 7 h 14 min
From midnight until 08 13 = 8 h 13 min
Total length of = 7 h 14 min + 8 h 13 min
= 15 h 27 min

## Exercise 1

**1** There are many different units to measure time (e.g. seconds, hours, months).
Choose the most sensible unit of time to measure each of these activities.
a) time taken to bake a cake
b) length of time for the school holidays in summer
c) time taken to travel by plane from England to New Zealand
d) time taken to run 100 metres
e) time taken to boil an egg
f) length of time most people live
g) time taken to play a football match
h) time taken to blink your eyelid
i) time taken to build a house
j) length of time a person sleeps in 1 day

**2** Calculate the number of minutes in each of these times.
a) $\frac{1}{2}$ hour                                b) $\frac{3}{4}$ hour
c) $2\frac{1}{3}$ hours                              d) $4\frac{1}{4}$ hours
e) $5\frac{5}{6}$ hours                              f) $7\frac{2}{3}$ hours
g) $15\frac{1}{4}$ hours                             h) $20\frac{1}{6}$ hours
i) 1 day and 4 hours

**3** Calculate the number of hours in each of these times.

a) $\frac{1}{2}$ day

b) $\frac{3}{4}$ day

c) $2\frac{1}{3}$ days

d) $4\frac{1}{4}$ days

e) $5\frac{5}{6}$ days

f) $7\frac{2}{3}$ days

g) 1 week and 3 days

h) 1 week and $4\frac{1}{2}$ days

i) 2 weeks and $2\frac{1}{4}$ days

**4** Convert (change) each of these times to the units given in brackets.

a) 5 days (hours)

b) 3 hours (minutes)

c) 6 hours (minutes)

d) 240 seconds (minutes)

e) 96 hours (days)

f) 360 minutes (hours)

g) 250 minutes (hours)

h) 25 minutes (seconds)

i) 100 hours (days)

j) 220 minutes (hours)

k) $3\frac{1}{2}$ days (hours)

l) $2\frac{3}{4}$ hours (minutes)

**5** Add up these times.

a) 4 h 7 min + 2 h 24 min

b) 6 h 37 min + 5 h 12 min

c) 1 h 48 min + 2 h 8 min

d) 3 h 23 min + 4 h 7 min

e) 7 h 12 min + 3 h 52 min

f) 2 h 39 min + 8 h 43 min

g) 6 h 56 min + 2 h 31 min

h) 3 h 19 min + 1 h 46 min

i) 2 h 42 min + 2 h 17 min

**6** Rewrite each of these times using the 24-hour clock.

a) 8.50 a.m.

b) 11.22 a.m.

c) 12.42 a.m.

d) 1.05 a.m.

e) 2.53 a.m.

f) 4.15 p.m.

g) 7.38 p.m.

h) 3.12 p.m.

i) 9.43 p.m.

j) 11.50 p.m.

**7** Rewrite each of these times using the 12-hour clock (with a.m. or p.m.).

a) 02 30

b) 07 40

c) 00 15

d) 11 12

e) 12 00

f) 19 00

g) 13 42

h) 20 20

i) 22 32

j) 23 10

**8** Calculate how much time is between each of these two times (in the same day).

a) 06 00 and 16 00

b) 03 00 and 23 00

c) 08 30 and 19 30

d) 12 50 and 21 50

e) 06 40 and 23 45

f) 8.00 a.m. and 2.00 p.m.

g) 4.00 a.m. and 7.00 p.m.

h) 1.30 a.m. and 12.30 p.m.

i) 11.45 a.m. and 10.45 p.m.

9 In each pair of times, the first time is 'today' and the second time is 'tomorrow'. Calculate how much time is between each of the two times.

a) 23 00 and 05 00
b) 18 00 and 10 00
c) 14 15 and 08 15
d) 21 45 and 03 30
e) 22 28 and 20 44
f) 7.00 p.m. and 3.00 a.m.
g) 2.00 p.m. and 9.00 a.m.
h) 6.20 p.m. and 3.20 a.m.
i) 8.06 p.m. and 4.24 p.m.

10 Here are the times of some UK TV programmes on Sunday evening.

| 6.35 | 7.20 | 8.00 | 8.50 |
|---|---|---|---|
| Emmerdale | Coronation Street | Heartbeat | News |

a) Write the times of the programmes using the 24-hour clock.
b) For how many minutes does *Coronation Street* last?
c) For how many minutes does *Emmerdale* last?

11 A train left Paddington station, London at 1.15 p.m. and arrived in Exeter at 4.05 p.m.

a) Write the times using the 24-hour clock.
b) How long did the journey take?

12 A train left Toronto at 11.10 a.m. and arrived in Montreal at 3.25 p.m.

a) Write the times using the 24-hour clock.
b) How long did the journey take?

13 A bus leaves Phuket at 10.50 p.m. to travel to Krabi. The journey takes 2 hours 40 minutes. At what time does the bus arrive in Krabi?
Give your answer in
a) 24-hour clock time
b) 12-hour clock time.

14 Mrs Hill took 3 hours 56 minutes to drive from Brussels to Frankfurt. She left Brussels at 10 45. At what time did she arrive in Frankfurt?
Give your answer in
a) 24-hour clock time
b) 12-hour clock time.

15 Mr Stevens works for $4\frac{1}{4}$ hours each morning and for $3\frac{3}{4}$ hours each afternoon. He stops work for lunch from 12 30 until 13 45.
At what time each day does he
a) start work
b) finish work?

## C   Length

In the metric system of measures, the basic unit used to measure length is the **metre** (m). In fact, the metric system gets its name from this unit of length.

All the other units used to measure length in the metric system are connected to the metre.

### Metric units of length

| | |
|---|---|
| 1 metre | = 10 **decimetres** (dm) |
| 1 metre | = 100 **centimetres** (cm) |
| 1 metre | = 1000 **millimetres** (mm) |
| 1 decimetre | = $\frac{1}{10}$ metre |
| 1 centimetre | = $\frac{1}{100}$ metre |
| 1 millimetre | = $\frac{1}{1000}$ metre |
| 1000 metres | = 1 **kilometre** (km) |
| 1 metres | = $\frac{1}{1000}$ kilometre |

NOTE: We don't use the unit **decimetre** very much – most measurements are made with metre, centimetre and millimetre.

We can use the following approximations to help us decide which units are most sensible for different lengths.

- 1 metre is about the length from the left side of your body to the tip of your right hand.
- 1 centimetre is about the width of your little finger.
- 1 millimetre is about the thickness of 20 pages of a book.

If we wanted to measure the width of this book, it would be silly to use metres. The book is not nearly as long as your arm! Centimetres would be much more sensible.

Each unit of length is related to the others by some power of 10. This means that to convert (change) one unit of length into another, we must multiply or divide by some power of 10.

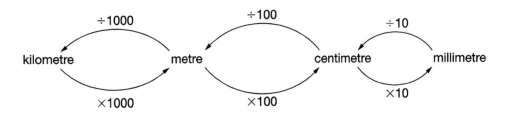

To convert (change) from a **larger** measurement to a **smaller** measurement, we **multiply** by a power of 10.

**Examples**

Convert each of these measurements to the unit shown in brackets.

**a)** 6 cm (mm)

$1\,\text{cm} = 10\,\text{mm}$
$6\,\text{cm} = 6 \times 10\,\text{mm}$
$\qquad = 60\,\text{mm}$

**b)** 4.25 km (m)

$1\,\text{km} \quad = 1000\,\text{m}$
$4.25\,\text{km} = 4.25 \times 1000\,\text{m}$
$\qquad\quad = 4250\,\text{m}$

**c)** $3\frac{1}{2}$ m (cm)

$1\,\text{m} \quad = 100\,\text{cm}$
$3\frac{1}{2}\,\text{m} \;=\; 3\frac{1}{2} \times 100\,\text{cm}$
$\qquad = \frac{7}{2} \times 100\,\text{cm}$
$\qquad = 350\,\text{cm}$

To convert (change) from a **smaller** measurement to a **larger** measurement, we **divide** by a power of 10.

**Examples**

Convert each of these measurements to the unit shown in brackets.

**a)** 500 m (km)

$1\,\text{m} \quad = \frac{1}{1000}\,\text{km}$
$500\,\text{m} \;=\; 500 \div 1000\,\text{km}$
$\qquad = 0.5\,\text{km}$

**b)** 35.6 mm (cm)

$1\,\text{mm} \quad = \frac{1}{10}\,\text{cm}$
$35.6\,\text{mm} \;=\; 35.6 \div 10\,\text{cm}$
$\qquad\quad = 3.56\,\text{cm}$

**c)** 350 cm (m)

$1\,\text{cm} \quad = \frac{1}{100}\,\text{m}$
$350\,\text{cm} \;=\; 350 \div 100\,\text{m}$
$\qquad\quad = 3.5\,\text{m}$

Sometimes we need to multiply or divide **more than once**.

**Examples**

**a)** Convert 2 km to centimetres.

$1\,\text{km} \quad = 1000\,\text{m} \text{ and } 1\,\text{m} = 100\,\text{cm}$
$2\,\text{km} \quad = 2 \times 1000\,\text{m}$
$\qquad = 2000\,\text{m}$
$2000\,\text{m} \;= 2000 \times 100\,\text{cm}$
$\qquad\quad = 200\,000\,\text{cm}$
So $2\,\text{km} = 200\,000\,\text{cm}$

**b)** Convert 150 000 mm to kilometres.

$1\,mm = \frac{1}{10}\,cm$, $1\,cm = \frac{1}{100}\,m$ and $1\,m = \frac{1}{1000}\,km$

| 150 000 mm | = 150 000 ÷ 10 cm |
| | = 15 000 cm |
| 15 000 cm | = 15 000 ÷ 100 m |
| | = 150 m |
| 150 m | = 150 ÷ 1000 km |
| | = 0.15 km |

So 150 000 mm = 0.15 km

## Exercise 2

**1** Decide which unit of length (km, m, cm or mm) you would use to measure each of these.
  a) height of the classroom wall
  b) thickness of a pencil point
  c) width of a credit card
  d) distance between Paris and Brussels
  e) width of a blade of grass
  f) length of your foot
  g) width of Ireland
  h) distance from your home to school

**2** To convert one unit of length to another, we must multiply or divide by some power of 10. Write down what you would use to convert each of these units.
  a) cm to mm
  b) cm to m
  c) m to mm
  d) mm to m
  e) km to cm
  f) m to km
  g) cm to km
  h) m to cm

**3** Copy and complete each calculation. Write a number in each space to make the calculation true. The first one is done for you.
  a) 2 m = __2__ × 100 = __200__ cm
  b) 6 m = _____ × 1000 = _____ mm
  c) 850 cm = _____ ÷ 100 = _____ m
  d) 4.5 cm = _____ × 10 = _____ mm
  e) 1.34 km = _____ × 1000 = _____ m
  f) 72.6 mm = _____ ÷ 1000 = _____ m
  g) 12.6 m = _____ × 100 = _____ cm
  h) 4820 m = _____ ÷ 1000 = _____ km

**4** Copy and complete each calculation. Write a number in each space to make the calculation true. The first one is done for you.
  a) 12 m = 12 × __100__ = __1200__ cm
  b) 45 m = 45 × _____ = _____ mm
  c) 250 m = 250 ÷ _____ = _____ km
  d) 74 mm = 74 ÷ _____ = _____ cm
  e) 2.2 km = 2.2 × _____ = _____ m
  f) 38 cm = 38 × _____ = _____ mm
  g) 450 mm = 450 ÷ _____ = _____ cm
  h) 305 m = 305 ÷ _____ = _____ km

**5** Convert each of these measurements to metres.
  a) 3 km
  b) 5.6 km
  c) 6.34 km
  d) $7\frac{1}{2}$ km

6 Convert each of these measurements to centimetres.
   a) 11 m
   b) 0.63 m
   c) 3.59 m
   d) $13\frac{1}{2}$ m

7 Convert each of these measurements to millimetres.
   a) 35 cm
   b) 0.52 cm
   c) 5.83 cm
   d) $5\frac{1}{2}$ cm

8 Convert each of these measurements to millimetres.
   a) 50 m
   b) 0.07 m
   c) 4.321 m
   d) $3\frac{3}{4}$ m

9 Convert each of these lengths to the unit given in brackets.
   a) 3 m (cm)
   b) 4 km (m)
   c) 2 m (mm)
   d) 10 cm (mm)
   e) 4.9 km (m)
   f) 5.13 m (mm)
   g) 68.2 m (cm)
   h) 5.24 cm (mm)
   i) 37.1 m (cm)
   j) $3\frac{1}{10}$ km (m)
   k) $6\frac{1}{2}$ cm (mm)
   l) $2\frac{7}{10}$ m (mm)

10 Convert each of these lengths to the unit given in brackets.
   a) 6000 cm (mm)
   b) 50 mm (cm)
   c) 5500 mm (m)
   d) 682 m (km)
   e) 482 cm (km)
   f) 926 mm (m)
   g) 43 m (mm)
   h) 60 km (cm)
   i) 0.141 km (mm)

## D Mass

We often talk about 'mass' and 'weight' as if they are the same thing. In fact, in scientific terms they are not exactly the same. You will learn more about this in science lessons. For now, all you need to know is that when you **weigh** something, you are actually finding its **mass**!

In the metric system of measures, the basic unit used to measure mass is the gram (g).

All the other units used to measure mass in the metric system are connected to the gram.

## Metric units of mass

| | |
|---|---|
| 1 gram | $= 1000$ **milligrams** (mg) |
| 1 milligram | $= \frac{1}{1000}$ gram |
| 1000 grams | $= 1$ **kilogram** (kg) |
| 1 gram | $= \frac{1}{1000}$ kg |
| 1 milligram | $= \frac{1}{1\,000\,000}$ kg |
| 1000 kilogram | $= 1$ **tonne** (t) |
| 1 kilogram | $= \frac{1}{1000}$ tonne |
| 1 gram | $= \frac{1}{1\,000\,000}$ tonne |
| 1 milligram | $= \frac{1}{1\,000\,000\,000}$ tonne |

We can use the following approximations to help us decide which units are most sensible for different lengths.

- 1 gram is about the mass of 1 drawing pin.
- The mass of a headache tablet is about 500 milligrams.
- 1 kilogram is about the mass of a bag of sugar.
- 1 tonne is about the mass of a motorcar.

Each unit of mass is related to the others by some power of 10. This means that to convert (change) one unit of mass into another, we must multiply or divide by some power of 10.

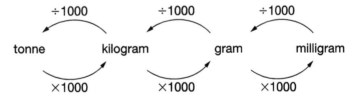

**Examples**

Convert each of these masses to the unit given in brackets.

**a)** 3t (kg)

$1\,t = 1000\,kg$

$3\,t = 3 \times 1000\,kg$

$\phantom{3\,t} = 3000\,kg$

**b)** $4\frac{1}{2}$ kg (g)

$1\,kg = 1000\,g$

$4\frac{1}{2}\,kg = \frac{9}{2} \times 1000\,g$

$\phantom{4\frac{1}{2}\,kg} = 4500\,g$

**c)** 375 g (kg)

$1\,g = \frac{1}{1000}\,kg$

$375\,g = 375 \div 1000\,kg$

$\phantom{375\,g} = 0.375\,kg$

**d)** 1500 kg (t)

$1\,kg = \frac{1}{1000}\,t$

$1500\,kg = 1500 \div 1000\,t$

$\phantom{1500\,kg} = 1.5\,t$

**e)** 9.6 g (mg)

$1\,g = 1000\,mg$

$9.6\,g = 9.6 \times 1000\,mg$

$\phantom{9.6\,g} = 9600\,mg$

**f)** 8120 mg (g)

$1\,mg = \frac{1}{1000}\,g$

$8120\,mg = 8120 \div 1000\,g$

$\phantom{8120\,mg} = 8.12\,g$

**Exercise 3**

1  Decide which unit of mass (mg, g, kg or tonne) you would use to measure each of these.
   a) the mass of a bag of apples        b) the mass of a person
   c) the mass of a ship               d) the mass of a box of matches
   e) the mass of a chair              f) the mass of a train
   g) the mass of Vitamin C in 1 orange  h) the mass of calcium in 1 glass of milk

2  To convert one unit of mass to another, we must multiply or divide by some power of 10. Write down what you would use to convert each of these units.
   a) g to mg                         b) g to kg
   c) kg to mg                        d) mg to kg
   e) tonne to g                      f) kg to tonne
   g) g to tonne                      h) kg to g

3  Convert each of these masses to grams.
   a) 0.205 kg                        b) $2\frac{3}{10}$ kg
   c) 0.4875 kg                       d) 9.027 kg

4  Convert each of these masses to kilograms.
   a) 6 g                             b) 89 g
   c) 268.7 g                         d) 16.32 g

5  Convert each of these masses to kilograms.
   a) 3.54 t                          b) 0.8 t
   c) 12.017 t                        d) 23.704 t

6  Convert each of these masses to tonnes.
   a) 8 kg                            b) 560 kg
   c) 9101 kg                         d) 678.9 kg

7  Convert each of these masses to milligrams.
   a) $2\frac{3}{4}$ g                b) 0.5 g
   c) 0.73 g                          d) 13.58 g

8  Convert each of these masses to grams.
   a) 3 kg                            b) 5.6 kg
   c) 6.34 kg                         d) $7\frac{1}{2}$ kg

9  Convert each of these masses to grams.
   a) 640 mg                          b) 60 mg
   c) 7 mg                            d) 9000 mg

10 Convert each of these masses to the unit given in brackets.
a) 6.2 kg (g)                      b) 4560 kg (t)
c) $5\frac{1}{2}$ kg (g)                   d) 470 t (kg)
e) 609 kg (t)                      f) 2.25 t (kg)
g) $5\frac{1}{2}$ t (kg)                    h) 68 g (kg)
i) 5280 mg (g)                     j) 9.06 g (mg)
k) 963 g (t)                       l) 0.000 74 t (g)

11 Mrs Cooper's cat weighed 4.3 kg in April. By July, its mass had increased by 543 g. What is the new mass of the cat in July? Give your answer in kilograms.

12 1 teaspoon of milk powder contains 23.4 mg of calcium. What is the mass of calcium contained in 100 teaspoons of milk powder? Give your answer in grams.

13 An empty truck weighs 3.4 tonnes. It carries some parcels that weigh 796 kilograms. What is the total mass of the truck and the parcels? Give your answer in tonnes.

14 Flour is packed in bags of 2.5 kg each. I have 1 tonne of flour. How many bags of flour can I pack?

## E Volume (or capacity)

The word 'capacity' means 'how much we can put inside something'.

The word 'volume' means 'how much liquid or gas we have' or 'how much space is inside something'.

You can see that they are very similar words, but 'capacity' is used more often to describe the space inside a container, and volume is more often used to describe a quantity of liquid or gas.

In the metric system of measures, the basic unit used to measure volume (or capacity) is the litre (l).

There are many other units used to measure volume (or capacity) in the metric system; they are all connected to the litre. For now, though, we will learn about only one other unit – the millilitre (ml).

# Metric units of volume (or capacity)

1 litre = 1000 millilitres (ml)
1 millilitre = $\frac{1}{1000}$ litre

NOTE: The abbreviation for 'litre' is 'l', but we usually write the word out in full, because 'l' looks very like the number 1, which can be confusing.

The conversions between litres and millilitres work like this.

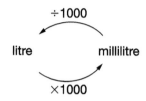

**Examples** Convert each of these volumes to the unit given in brackets.

a) $1\frac{1}{2}$ litres (ml)

1 litre = 1000 ml
$1\frac{1}{2}$ litres = $\frac{3}{2} \times 1000$ ml
= 1500 ml

b) 500 ml (litres)

1 ml = $\frac{1}{1000}$ litre
500 ml = 500 ÷ 1000 litres
= 0.5 litres

## Exercise 4

1 Decide which unit of volume (ml or litre) you would use to measure each of these.
a) the capacity of a swimming pool
b) the capacity of a teacup
c) the amount of petrol in a can
d) the amount of beer in a bottle
e) the amount of medicine you should take
f) the amount of orange juice in a bottle
g) the amount of fish sauce you put on your meal
h) the amount of water to put out a big fire

2 Convert each of these volumes to the units given in brackets.
a) 75 000 ml (litres)
b) 69 litre (ml)
c) 4.53 litre (ml)
d) 13.87 litre (ml)
e) 5350 ml (litres)
f) 4580 ml (litres)
g) $5\frac{1}{2}$ litres (ml)
h) $7\frac{1}{4}$ litres (ml)
i) 28 ml (litres)

3  Add each of these sets of volumes. Write your answer in the units given
   in brackets.
   a) 34 litres + 75 litres (ml)          b) 6.8 litres + 550 ml (ml)
   c) 13.5 litres + 780 ml (ml)           d) 20 litres + 13.5 litres + 173 ml (ml)
   e) 73 litres + 23 500 ml (litres)      f) 54.8 litres + 17 litres + 9750 ml (litres)
   g) 35 litres + 19 500 ml (litres)      h) 23.6 litres + 32 litres + 3450 ml (litres)

4  Some pink paint is made by mixing half a litre of white paint with
   25 millilitres of red paint. How much paint is there altogether?
   Give your answer in millilitres.

5  Candy made a milkshake by mixing 0.25 litres of milk with 24 ml of
   chocolate sauce. What is the total volume of the milkshake?
   Give your answer in millilitres.

6  Mohinder had 1 litre of orange juice. He drank 225 ml of the juice.
   How many millilitres of juice are left over?

##  Estimating with sensible units and degrees of accuracy

It is very useful to be able to estimate length, mass and volume: it
means that we can tell the approximate size of something before we
measure it exactly.

Length
● Most adults are between 1.5 m and 1.8 m tall.
● The door to the classroom is about 2 m high.

Find some other facts that will help you to estimate length and
distance.

Mass
● A pen weighs about 5 g.
● A normal bag of sugar weighs 1 kg.

Find some other facts that will help you to estimate mass.

Volume
● A tablespoon holds about 15 ml.
● A can of soda holds about 330 ml.

Find some other facts that will help you to estimate volume and
capacity.

**Exercise 5**

1 Which of these is the best estimate for the mass of a banana?
   a) 1 kg                          b) 5 g
   c) 250 g                         d) 30 g
   e) 3 kg                          f) 750 g

2 Which of these is the best estimate for the diameter of a football?
   a) 2 m                           b) 50 mm
   c) 30 cm                         d) 1.5 m
   e) 0.6 m                         f) 800 mm

3 Which of these is the best estimate for the capacity of a mug?
   a) 15 ml                         b) 1200 ml
   c) 2 litre                       d) 0.5 litre
   e) 200 ml                        f) 800 ml

4 Give a sensible estimate, using the most suitable unit, for each
   of these measures.
   a) the length of a matchstick     b) the length of a football field
   c) the mass of a 30 cm plastic ruler   d) the mass of a bus

5 This picture shows a man standing next to a tree.

   Use something in the picture that you know the
   approximate size of to estimate the height of the tree.
   Choose a sensible metric unit for your answer.

6 This picture shows a large house.

   Use something in the picture that you
   know the approximate size of to estimate
   how wide the house is. Choose a sensible
   metric unit for your answer.

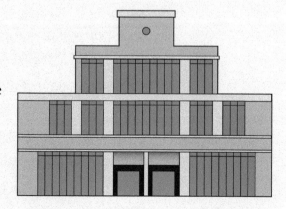

# Unit 9 An introduction to algebra

## Key vocabulary

| | | |
|---|---|---|
| addition | evaluate | simplify |
| algebra | expand | substitute |
| algebraic expression | expansion | subtraction |
| bracket | expression | sum |
| coefficient | like terms | symbol |
| constant | minus sign | term |
| constant term | multiplication | unlike terms |
| difference | plus sign | variable |
| distributive law | product | variable term |
| division | quotient | |

## A Mathematical operations

So far, we have learned to work out calculations with numbers, using the four basic mathematical operations.

| | |
|---|---|
| addition (+) | The answer to an addition calculation is called the sum. |
| subtraction (−) | The answer to a subtraction calculation is called the difference. |
| multiplication (×) | The answer to a multiplication calculation is called the product. |
| division (÷) | The answer to a division calculation is called the quotient. |

On the next page, we can see some of the other words that we use to talk about these four operations.

| Operation | Other words | |
|---|---|---|
| addition | increased by | greater than |
| | more than | larger than |
| | total of | grow |
| | combined/together | increase |
| | sum | rise |
| | plus | gain |
| subtraction | decreased by | smaller than |
| | less/fewer than | decrease |
| | difference between | lower |
| | minus | reduce |
| | less | lose |
| multiplication | times | twice |
| | of | double |
| | product of | triple |
| division | per | half |
| | out of | third |
| | ratio of | quarter |
| | quotient | . . . etc. |
| equals | is | will be |
| | are | gives |
| | was | yields |
| | were | sold for |

## B Algebraic expressions

Look at this sentence.

Twenty-three multiplied by eleven equals two hundred and fifty-three. If we had to do all our calculations in maths using all these words, it would take a very long time, and be very difficult!

We already know how to change a word sentence like this into a 'maths sentence' using the symbols for the numbers and the operations: $23 \times 11 = 253$. Look at this sentence.

Thirteen times **a number** equals three hundred and fifty-one.

In this calculation, we do not know what all the numbers are.

We can write this as a maths sentence by using a symbol to mean 'a number'. We could use any symbol we like, but it is usual to choose a letter of the alphabet – often the first letter of the word. For this calculation, we could use the letter $n$ as a symbol for 'a number': $13 \times n = 351$.

We can do the same with more complicated calculations.

**Example**

Write this sentence using symbols.
The sum of twice a number and three times the same number, equals five times that number.

Look at each part of the sentence separately.

| | |
|---|---|
| The sum of … and | This tells us that the calculation is an addition. |
| twice a number | We write this as $(2 \times n)$ |
| three times the same number | We write this as $(3 \times n)$ |
| equals | This tells us to write an equals sign. |
| five times that number | We write this as $(5 \times n)$ |

So we can write the whole sentence in symbols as
$(2 \times n) + (3 \times n) = (5 \times n)$

We can 'translate' **any** calculation into the 'language' of maths using symbols. This 'language' is called algebra.

The word we use to describe a 'maths sentence' is 'expression'. Because an expression is part of the 'language' of algebra, the full name of an expression is 'algebraic expression'.

As in the example above, we usually work through the sentence, translating each part. If the sentence says 'add' (or another term that means 'add', such as 'sum'), we write '+'. When we are multiplying, however, we don't always write the multiplication sign ($\times$). Use these rules to help you decide whether to write a multiplication sign.

- If an expression includes a number multiplied by a letter (in place of an unknown number), we usually **do not** write the multiplication sign.

For example, we write $2 \times b$ as $2b$.

- If the number is 1, we usually just write the letter.

For example, we write $1 \times d$ as $d$.

- If two (or more) letters (for unknown numbers) are multiplied together, we usually **do not** write the multiplication symbol.

For example, we write $a \times b$ as $ab$.

- If a number or a letter (or a combination of both) and the contents of brackets are multiplied together, we usually **do not** write the multiplication sign.

For example, we write $2 \times (a + b)$ as $2(a + b)$.

- If two numbers are multiplied together, we **must always** write the multiplication symbol.

For example, $7 \times 4$ must always be written as $7 \times 4$. We cannot write 74!

Here are some examples of algebraic expressions.

| Sentence | Algebraic expression |
|---|---|
| Add a number $a$ to three. | $3 + a$ |
| Add three to a number $a$. | $a + 3$ |
| Multiply two by a number $b$. | $2 \times b$ or $2b$ |
| The sum of $b$ and $c$. | $b + c$ |
| Subtract four from $c$. | $c - 4$ |
| Five times $d$. | $5 \times d$ or $5d$ |
| $e$ multiplied by six. | $e \times 6$ or $6e$ |
| The product of $x$ and $y$. | $x \times y$ or $xy$ |
| Divide seven by $r$. | $7 \div r$ or $\frac{7}{r}$ |
| One third of $h$. | $\frac{1}{3} \times h$ or $\frac{h}{3}$ |
| Subtract two from the sum of $a$ and $b$. | $(a + b) - 2$ |
| Add three to the product of $c$ and $d$. | $cd + 3$ |
| The quotient of $x$ and the product of $y$ and $z$. | $\frac{x}{yz}$ |
| Divide the sum of $r$ and $s$ by $t$. | $\frac{r+s}{t}$ |
| The product of $k$ with the sum of $m$ and $n$. | $k(m + n)$ |

## Exercise 1

**1** Choose the correct algebraic expression for each of the sentences.

a) 3 times a number                    i) $x + 3$
b) 3 more than a number                ii) $3x - 8$
c) 3 decreased by a number             iii) $x - 3$
d) 3 less than a number                iv) $3x + 8$
e) one third of a number               v) $3x$
f) 8 more than 3 times a number        vi) $3 - x$
g) 8 less than 3 times a number        vii) $\frac{x}{3}$

**2** Choose the correct algebraic expression for each of the sentences.

a) 7 less than 4 times a number          i) $7 - 4x$
b) 7 decreased by 4 times a number       ii) $2x - 9$

c) 9 less than twice a number

d) 9 decreased by twice a number

e) 9 less than half a number

f) 7 times a number, increased by 4

g) 7 times a number, increased by 4 times the number

iii) $7x + 4$

iv) $4x - 7$

v) $7x + 4x$

vi) $9 - 2x$

vii) $\frac{x}{2} - 9$

3 Choose the correct algebraic expression for each of the sentences.

a) 9 metres higher than the height $x$ metres

b) 15 metres per second slower than the speed $x$ m/s

c) 15 °C hotter than the temperature $x$ °C

d) 9 metres shorter than twice the length $x$ metres

e) 9 years older that twice the age $x$ years

f) $9 cheaper than 4 times the price $x

g) 9 cm less than three quarters of the length $x$ cm

i) $x + 15$

ii) $x + 9$

iii) $4x - 9$

iv) $2x - 9$

v) $2x + 9$

vi) $x - 15$

vii) $\frac{3}{4}x - 9$

4 Write each of these sentences as an algebraic expression.

a) Two times $c$.

b) The product of two and $d$.

c) Twice $e$.

d) Subtract three from $f$.

e) Divide three by $y$.

f) Add five to the product of $h$ and $k$.

g) Subtract the product of $m$ and $n$ from six.

h) Divide the sum of $p$ and $q$ by $r$.

i) Divide the product of $s$ and $t$ by $u$.

j) Subtract one quarter of $x$ from $y$.

k) Subtract $a$ from $b$ and multiply the result by $c$.

l) Add the product of $h$ and $k$ to the product of $x$ and $y$.

m) $a$ is increased by twice $b$.

n) Twice the sum of $a$ and $b$.

o) Thirty decreased by three times $c$.

p) Three times the difference of thirty and $c$.

q) Fifty minus the product of ten and $p$.

r) The product of fifty and the sum of $p$ and ten.

s) One hundred increased by the quotient of $x$ and $y$.

t) The quotient of $x$, and the sum of $y$ and one hundred.

u) Half of $a$ increased by the product of twenty-five and $b$.

v) Four times $c$ decreased by one-fifth of $d$.

w) Half the sum of $m$ and twice $n$.

x) Sixty reduced by one-third the product of seven and $x$.

# C Evaluating algebraic expressions

Now we know how to write an algebraic expression using numbers, symbols, and **letters** (that stand for other numbers that we don't yet know).

When we **replace** the letter in an expression with a number, the value of the expression will depend on the value of the number(s). If we use different numbers, the expression will have a different value for each of the numbers.

For example, look at the expression $3b + 4$.

If we let $b = 2$ and we substitute this number in place of $b$ in the expression, we get

$$3b + 4 = 3 \times 2 + 4$$
$$= 6 + 4$$
$$= 10$$

When $b = 2$, the **value** of the expression $3b + 4$ is 10.

We call finding the value of an expression evaluating the expression.

Let's look at the value of our expression $3b + 4$ when we replace $b$ with other numbers.

Let $b = 12$

$$3b + 4 = 3 \times 12 + 4$$
$$= 36 + 4$$
$$= 40$$

Let $b = 111$

$$3b + 4 = 3 \times 111 + 4$$
$$= 333 + 4$$
$$= 337$$

Let $b = \frac{1}{2}$

$$3b + 4 = 3 \times \frac{1}{2} + 4$$
$$= 1\frac{1}{2} + 4$$
$$= 5\frac{1}{2}$$

Let $b = -5$

$$3b + 4 = 3 \times (-5) + 4$$
$$= -15 + 4$$
$$= -11$$

You can see that the value of the expression changes with each value we give to the letter $b$.

Sometimes an expression involves more than one letter. We replace each letter with a number in just the same way.

**Examples**

If $b = 2$, $k = 3$, $p = -1$, $q = -2$ and $r = 0$, evaluate each of these expressions.

a) $b + 5k$

$$b + 5k = 2 + (5 \times 3)$$
$$= 2 + 15$$
$$= 17$$

b) $3b + p$

$$3b + p = (3 \times 2) + (-1)$$
$$= 6 - 1$$
$$= 5$$

c) $4k - 3q$

$$4k - 3q = (4 \times 3) - [3 \times (-2)]$$
$$= 12 - (-6)$$
$$= 12 + 6$$
$$= 18$$

d) $9r + bk - pq$

$$9r + bk - pq = (9 \times 0) + (2 \times 3) - [(-1) \times (-2)]$$
$$= 0 + 6 - (+2)$$
$$= 6 - 2$$
$$= 4$$

Sometimes the expression we need to evaluate involves brackets. Remember that when numbers are multiplied by numbers we **must always** write the multiplication sign – so when we substitute numbers for letters, we must also put the multiplication sign(s) back in the expression.

**Examples**

If $u = 2$, $v = 3$, $w = 5$, $x = 0$ and $y = -1$, evaluate each of these expressions.

a) $u(v + w)$

$$u(v + w) = 2(3 + 5)$$
$$= 2 \times 8$$
$$= 16$$

b) $v(w - x)$

$$v(w - x) = 3(5 - 0)$$
$$= 3 \times 5$$
$$= 15$$

c) $y(uv - w)$

$$y(uv - w) = (-1)[(2 \times 3) - 5]$$
$$= (-1) \times (6 - 5)$$
$$= (-1) \times 1$$
$$= -1$$

d) $(2x - 3y)u$

$$(2x - 3y)u = [(2 \times 0) - (3 \times \{-1\})] \times 2$$
$$= [0 - (-3)] \times 2$$
$$= (0 + 3) \times 2$$
$$= 3 \times 2$$
$$= 6$$

We learned about power numbers in Unit 5. The rules are the same when we use letters instead of numbers. So $m^2 = m \times m$ and $n^3 = n \times n \times n$, and so on.

**Examples**

If $m = 4$ and $n = 6$, evaluate each of these expressions.

a) $m^2n$

$$m^2n = 4 \times 4 \times 6$$
$$= 16 \times 6$$
$$= 96$$

c) $\dfrac{n^3}{2m}$

$$\frac{n^3}{2m} = \frac{\overset{3}{\cancel{6}} \times \overset{3}{\cancel{6}} \times \overset{3}{\cancel{6}}}{\underset{1}{\cancel{2}} \times \underset{\underset{1}{\cancel{2}}}{\cancel{4}}}$$
$$= 3 \times 3 \times 3$$
$$= 27$$

b) $\dfrac{1}{2}mn^2$

$$\frac{1}{2}mn^2 = \frac{1}{\underset{1}{\cancel{2}}} \times \overset{2}{\cancel{4}} \times 6 \times 6$$
$$= 2 \times 36$$
$$= 72$$

**Exercise 2**

1 If $a = 5$, $b = 2$, $c = -1$, $d = 3$ and $e = 0$, evaluate each of these expressions.

a) $3a$

b) $11c$

c) $a + 5$

d) $e - 5$

e) $8 - b$

f) $3d + 2$

g) $8 - 2c$

h) $6d - 7e$

i) $a + 2b + 3c$

j) $5e - 6a + 7b$

k) $7b - 8c - 9d$

l) $ab + ac$

m) $a(b + c)$

n) $ad - bd$

o) $d(a - b)$

p) $a^2b$

q) $\dfrac{1}{5}a^2c$

r) $\dfrac{ab}{2d^2}$

s) $a + (b^3 - c^2)$

t) $\dfrac{a^2 + b^2}{c^3 - d^2}$

2 Evaluate each of these expressions using the values given.

a) $4x + 12y$ when $x = 9$ and $y = 7$

b) $p^3 + q^2$ when $p = 2$ and $q = 5$

c) $3y^2$ when $y = -2$

d) $5p - 3q$ when $p = 4$ and $q = -2$

e) $2r^2 + rs$ when $r = 3$ and $s = 2$

f) $\dfrac{1}{2}c(a + b)$ when $c = 4$, $a = 5$ and $b = 7$

## D Terms, variables, coefficients and constants

In this new language of algebra, we have learned that a maths 'sentence' such as $\frac{a}{2} + b^2 - 3c + 5$ is called an **expression**.

In any language, sentences are made up of **words**. In algebra, maths 'sentences' are made in the same way. Each part of a maths sentence (expression) is like a maths 'word'. In algebra the name for each maths 'word' is 'term'.

There are four terms in the expression $\frac{a}{2} + b^2 - 3c + 5$:

- $\frac{a}{2}$ is one term
- $b^2$ is also a term
- $-3c$ is another term
- 5 is the last term in the expression.

NOTE: Each of the terms in an expression is joined by either a plus sign or a minus sign. We may not write the plus sign, but there **always** is a sign. Terms in an expression are always **added or subtracted**.

In English we make words by combining letters. In algebra, we can use both letters and numbers to make **terms**. Lets look at $\frac{a}{2} + b^2 - 3c + 5$ again.

- $\frac{a}{2} = \frac{1}{2}a = \frac{1}{2} \times a$    This term is made by multiplying a **number part**, $\frac{1}{2}$, and a **letter part**, $a$.

    or                This term is made by dividing a **letter part**, $a$, by a **number part**, 2.

- $b^2 = 1 \times b \times b$    This term is made by multiplying a **number part**, 1, and two **letter parts**, $b$ and $b$. (Remember: If the number part is 1, we do not normally need to write it in the expression.)

- $-3c$                This term is made by multiplying a **number part**, $-3$, and a **letter part**, $c$.

- 5                 This term is just a **number** on its own.

We know that the letters stand in place of **unknown numbers**, and that these numbers can have **different values**. So the value of the letters can **change**. Another word for 'change' is '**vary**', so we call the letter parts of a term variables. A term can have more than one variable.

If a number part is multiplied or divided by a letter part, this number part is called a coefficient. If there is a minus sign between two terms, the coefficient of the second term is negative.

If a number part is on its own with no letter part (variable), it is called a constant term ('constant' means 'unchanging').

Look at the following examples.

$5x^2 - 3ax + a$    In the term $5x^2$ the coefficient of $x^2$ is $+5$.
In the term $-3ax$ the coefficient of $ax$ is $-3$.
In the term $+a$ the coefficient of $a$ is $+1$.

$3a^2 + ab - 4 + 7a$    In the term $3a^2$ the coefficient of $a^2$ is $+3$.
In the term $+ab$ the coefficient of $ab$ is $+1$.
In the term $+7a$ the coefficient of $a$ is $+7$.
The term $-4$ is $a$ constant term (there is no variable).

We know that $\dfrac{1}{4} + \dfrac{2}{4} = \dfrac{1+2}{4}$.

Therefore we also know that $\dfrac{1+2}{4} = \dfrac{1}{4} + \dfrac{2}{4}$.

Therefore we can say that $\dfrac{ax+by}{3c} = \dfrac{ax}{3c} + \dfrac{by}{3c}$.

So the expression $\dfrac{ax+by}{3c}$ has two terms, not just one.

- $\dfrac{ax}{3c} = \dfrac{ax}{c} \times \dfrac{1}{3}$    The coefficient of $\dfrac{ax}{c}$ is $+\dfrac{1}{3}$.

- $\dfrac{by}{3c} = \dfrac{by}{c} \times \dfrac{1}{3}$    The coefficient of $\dfrac{by}{c}$ is also $+\dfrac{1}{3}$.

## E Like and unlike terms

Variable terms are made out of a **coefficient** (number part) and one or more **variables** (letter part).

- If two different terms have exactly the **same variables**, we call them like terms.
- If two different terms have **different variables**, we call them unlike terms.

Look at the following examples.

The terms $2x$ and $3x$ are **like terms**. They have exactly the same variable part ($x$) even though the coefficients (number parts) are different.

The terms $2x$ and $2x^2$ are **unlike terms**. They have different variable parts ($x$ and $x^2$), even though the coefficients (number parts) are exactly the same.

The terms $2xy$ and $3yx$ are **like terms**. They have exactly the same variable part ($xy = yx$) even though the coefficients (number parts) are different.

The terms $2ab$ and $ab^2$ are **unlike terms**. They have different variable parts ($ab = a \times b$ and $ab^2 = a \times b \times b$) and the coefficients are also different.

We need to be able to recognise like terms within an expression.

**Examples**

Write down the like terms in each of these expressions.
a) $m^2 + 5m^3 + 2m^2$
   There are **two like terms**: $m^2$ and $2m^2$.
b) $pq - 2qr + 3rs - 2qrs$
   There are **no like terms**: all four terms are **unlike**.

### Exercise 3

**1** Here are some algebraic terms. Write down whether each term is a constant or a variable term.
   a) $6$
   b) $x$
   c) $6x^2$
   d) $xy$
   e) $\frac{3}{5}xy^2$
   f) $7\frac{2}{3}$

**2** a) Write down the coefficient of $a^2$ in the expression $7a^2 + 1$.
   b) Write down the coefficient of $x$ in the expression $x^2 - 5x + 4$.
   c) Write down the coefficient of $by$ in the expression $3y^2 + by - 2b^2$.
   d) Write down the coefficient of $gh$ in the expression $5 + g^2 - 4gh + h^2$.

**3** a) Write down the constant term in the expression $7a^2 + 1$.

   b) Write down the constant term in the expression $x^2 - 5x + 4$.

   c) Write down the constant term in the expression $3y^2 + by - 2b^2$.

   d) Write down the constant term in the expression $5 + g^2 - 4gh + h^2$.

**4** Write down how many terms there are in each of these expressions.

   a) $3x - 5$          b) $x^2y$

   c) $5x^2 - 3ax + a$     d) $\dfrac{3a}{5} + 7x + 4$

   e) $\dfrac{x^2}{ay + 9}$        f) $3a^2 + ab - 4 + 7a$

   g) $5 + 5by - 2x$     h) $3c - 2x + 1$

   i) $\dfrac{x}{4}$            j) $\dfrac{ab - bx}{7}$

**5** Write down the like terms in each of these expressions.

   a) $5a + 3b - 2a$           b) $m^2 - 3m^2 + m$

   c) $5b^2 - 3bk + kb$       d) $3a^2 + ab - 4 + 7ba$

   e) $5r^2 + 5r - 2r^3$        f) $-e^2 - 2ef + 6fe$

   g) $2yb^2 + yb - 5b^2y$    h) $8x - 3y + 7z + 5$

   i) $17e + 16 - 4e - 3$     j) $4mn + 2n + 3n - nm$

   k) $\frac{1}{4}x - 3 + 2x^2 + 5$      l) $3d^2 - 2g - g + d^2$

## F   Simplifying algebraic expressions

We know that $3 + 3 + 3 + 3$ can be written in a simpler way as '4 threes', or $4 \times 3$. In the same way, $a + a + a + a$ can be written as 4 times $a$, or $4 \times a$ which we write as $4a$.

So if we see expressions such as $5b + 2b$ and $6y - 3y$, we can write them as

$$5b + 2b = (b + b + b + b + b) + (b + b)$$
$$= b + b + b + b + b + b + b = 7b$$
$$\text{and} \quad 6y - 3y = (y + y + y + y + y + y) - (y + y + y)$$
$$= y + y + y + y + y + y - y - y - y = 3y$$

This is the same as **adding or subtracting** the **coefficients** (number parts) of the like terms. However, we can only do this because in each example the terms have exactly the same **variable** (letter part) and so are **like terms**.

In the expression $2b + 3y$ there are **no like terms**, and so we **cannot simplify** it in any way.

Only **like terms** can be added or subtracted.

**Examples**

Simplify these expressions, if possible.

a) $7a + 4a + 6a$

$7a + 4a + 6a = 17a$     All three terms in this expression are like terms, so we can add all the coefficients.

b) $9b + 5b - 3b$

$9b + 5b - 3b = 14b - 3b$     All three terms in this expression are
$= 11b$     like terms, so we can add and subtract all the coefficients.

Sometimes the expression we want to simplify contains terms in more than one variable. We may still be able to simplify the expression, but our final answer will have more than one term.

To simplify an expression:
1. Find all the like terms in the expression.
2. Write the expression again so that each group of like terms is written together.
3. Add or subtract the coefficients of all the like terms.

**Examples**

Simplify these expressions, if possible.

a) $4x + y + 2x + 3y$

$$4x + y + 2x + 3y$$
$$= \underbrace{4x + 2x} + \underbrace{y + 3y}$$
$$= \qquad 6x \quad + \quad 4y$$

$4x$ and $2x$ are like terms.
$y$ and $3y$ are also like terms.
Rewrite the expression with like terms together, and add.

$6x + 4y$ cannot be simplified any further because they are **not** like terms.

b) $3x + 4 + 5x - 2x - 3$

$$3x + 4 + 5x - 2x - 3$$
$$= \underbrace{3x + 5x - 2x} + \underbrace{4 - 3}$$
$$= \qquad 6x \qquad + \quad 1$$

$3x$, $5x$ and $-2x$ are like terms.
$4$ and $-3$ are also like terms.
like terms.
Rewrite the expression with like terms together, then add and subtract.

$6x + 1$ cannot be simplified any further because they are **not** like terms.

Remember that it doesn't matter what order the variables are in.
For example, $2ba = 2ab$.

**Examples**   Simplify these expressions, if possible.
a) $ab + a + 2b + 2ba$

$ab + a + 2b + 2ba$

$\quad = ab + a + 2b + 2ab$

*ab* and *2ba* are like terms. Remember, $2ba = 2ab$. Rewrite the expression with like terms together, and add.

$\quad = ab + 2ab + a + 2b$

$\quad = \underbrace{3ab} + a + 2b$

$3ab + a + 2b$ cannot be simplified any further because the terms are **not** like terms.

b) $m - mn + 2n + 3mn$

$m - mn + 2n + 3mn$

$\quad = \underbrace{-mn + 3mn} + m + 2n$

Rewrite the expression with like terms together.

$\quad = 2mn + m + 2n$

Add and subtract.

$2mn + m + 2n$ cannot be simplified any further because the terms are **not** like terms.

We simplify expressions involving fractions in exactly the same way.

**Examples**   Simplify these expressions, if possible.

a) $\dfrac{1}{4}r + \dfrac{1}{6}r$

$\dfrac{1}{4}r + \dfrac{1}{6}r = \left(\dfrac{1}{4} + \dfrac{1}{6}\right)r$   Both terms in this expression are like terms,

$\quad = \left(\dfrac{3 + 2}{12}\right)r$   so we can add them.

$\quad = \dfrac{5}{12}r$

b) $p - \dfrac{1}{2}p + \dfrac{1}{3}p$

$p - \dfrac{1}{2}p + \dfrac{1}{3}p = \left(1 - \dfrac{1}{2} + \dfrac{1}{3}\right)p$   All three terms in this expression are like terms,

$\quad = \left(\dfrac{6 - 3 + 2}{6}\right)p$   so we can add and subtract them.

$\quad = \dfrac{5}{6}p$

Sometimes we are given an expression that cannot be simplified.

**Examples**   Simplify these expressions, if possible.
a) $5p - 1q$
These are unlike terms, so we cannot simplify this expression.
b) $x^2 + 2x - 1$
These are unlike terms, so we cannot simplify this expression.

## Exercise 4

**1** Write each of these expressions in a simpler form (1 term).
a) $c + c + c$
b) $x + x + x + x + x$
c) $t + t + t - t$
d) $d + d - d + d - d$
e) $2n + n$
f) $2y + 3y$
g) $5g + g + 4g$
h) $5z + 4z + z - 3z$
i) $-4j + 2j + 5j$
j) $9c - 2c - 3c$
k) $3x - x + 5x$
l) $12w - 7w - 4w$
m) $5d + 7d - 12d$
n) $-2y - 3y$
o) $3x - 8x$
p) $2a - 5a - 12a + a$
q) $3b + 5b - 4b + 2b$
r) $m - 2m - 3m$

**2** Simplify each of these expressions as much as you can.
a) $5x + 3x + y$
b) $w + 3v - v$
c) $2a + b - 3b$
d) $2x + 3y + 3x$
e) $5 + 7u - 2$
f) $p + 3q + q$
g) $3d - 5c - 2c$
h) $3y + 1 - y$
i) $-a + b + 2a$
j) $3m + n + m$
k) $5c + 4c - d$
l) $2x + y - x$
m) $-p + 4p + 3p$
n) $5 - 9k + 4k$
o) $2a - a + 3$
p) $3s - 5 + t - 2t - 4s$

**3** Simplify each of these expressions as much as you can.
a) $3a + 5a + 2b + b$
b) $p + 2q + 2p + q$
c) $m + 2m - n + 3n$
d) $2x + 3y - x - 5y$
e) $3x - x + 5y - 2y$
f) $2d + 5 - d - 2$
g) $3a - 5a + 2b + b$
h) $a - 2a + 7 + a$
i) $2a - b + 3b - a$
j) $-f + g - f - g$
k) $2v - w - 3w - v$
l) $7 - 2t - 9 - 3t$
m) $-p + 3q - 3p + q$
n) $5 - 9k - 4 + 2k$
o) $2c + d + 4 - c - 2d + 7$
p) $2m + 3n - 6 - 4m$

**4** Simplify each of these expressions as much as you can.

a) $xy + yx$

b) $3pq - qp$

c) $5ab - 2ba$

d) $3x^2 - x^2$

e) $5y^2 + 4y^2$

f) $a^2 + 5a^2 - 2a^2$

g) $10ev - 2e + 5ve$

h) $3gh + 2g + 5gh + 4h$

i) $4mn + 2n + 3n^2 - nm$

j) $15jp + 5j - 14p - 3jp$

k) $8k - 2kr + 4rk - 8k$

l) $8st - 3s + 6t - 7$

**5** Simplify each of these expressions as much as you can.

a) $\frac{1}{3}b - \frac{1}{6}b$

b) $2d - \frac{1}{5}d$

c) $f + \frac{1}{4}f - \frac{1}{3}f$

d) $\frac{1}{2}h - k + h$

e) $\frac{1}{4}x - \frac{1}{7}y - \frac{1}{8}x$

f) $\frac{1}{6}j^2 - \frac{1}{3}p - \frac{1}{9}j^2$

g) $\frac{1}{2}uv + \frac{2}{3}vu$

h) $\frac{3}{10}mn - \frac{1}{5}mn + m$

**6** i) Simplify each these expressions as much as you can.

ii) Then find the value of the expression if $a = 7$ and $b = 3$.

a) $a + 2b$

b) $4a + 3b + 2a + b$

c) $6a + 5b - 4a - 3b$

d) $8a - 6b - 4a - b$

e) $9a - 6b - 4a - 2b$

f) $11ab - 3b - 2ba$

## G Adding and subtracting algebraic expressions

If we want to add or subtract different algebraic expressions, we can do this by writing the calculation so that we get just one expression. Then we can simplify this new expression by adding or subtracting any **like terms** in exactly the same way that we have done before.

**Example**

Add the expression $2a + 3b - c$ to the expression $4a - 4b + 5c$.

This is the same as
$(2a + 3b - c) + (4a - 4b + 5c) = 2a + 3b - c + 4a - 4b + 5c$.
This is really just one expression now, so we can group like terms and combine them as before.
$2a$ and $4a$ are like terms. $3b$ and $-4b$ are like terms. $-c$ and $5c$ are like terms.

$$2a + 3b - c + 4a - 4b + 5c = \underbrace{2a + 4a}_{6a} + \underbrace{3b - 4b}_{-b} \underbrace{- c + 5c}_{+ 4c}$$

The '−' sign in front of brackets tells us to subtract **each term** inside the bracket.

**Examples**

a) Subtract the expression $3x + 5y - 2z$ from the expression $7x - 2y - 6z$.

This is the same as

$(7x - 2y - 6z) - (3x + 5y - 2z) = 7x - 2y - 6z - 3x - 5y + 2z$

Notice how the signs of the final three expressions have changed.
We learned about adding and subtracting integers in Unit 1.
We do exactly the same with algebraic terms.
Now we have just one expression, so we can group like terms and combine them as before.

$$7x - 2y - 6z - 3x - 5y + 2z = 7x - 3x - 2y - 5y - 6z + 2z$$

$$= \quad 4x \quad\quad -7y \quad\quad -4z$$

b) Simplify this calculation as much as possible.

$(14x^2 + 3xy - y) + (-3x^2 + 5xy + 4y) - (5x^2 - xy + 9y)$

First we must remove the brackets to make one expression.
Be very careful with the signs!

$(14x^2 + 3xy - y) + (-3x^2 + 5xy + 4y) - (5x^2 - xy + 9y)$

$$= 14x^2 + 3xy - y - 3x^2 + 5xy + 4y - 5x^2 + xy - 9y$$

$$= 14x^2 - 3x^2 - 5x^2 + 3xy + 5xy + xy - y + 4y - 9y$$

$$= \quad 6x^2 \quad\quad\quad +9xy \quad\quad\quad -6y$$

**Exercise 5**

**1** Simplify each of these calculations as much as you can.

a) $(5a + 3b) + (a - 2b) + (3a + 5b)$

b) $(8x - 3y + 7z) + (-4x + 5y - 4z) + (-x - y - 2z)$

c) $(3b - 7c + 10) + (5c - 2b - 15) + (15 + 12c + b)$

d) $(9x - 6y - 7z) + (5x + 2y) + (3z + 5y + x)$

e) $(a - 3b + 3) + (2a + 5 - 3c) + (6c - 15 + 6b)$

f) $(13ab - 9cd - xy) + 5xy + (15cd - 7ab) + (6xy - 3cd)$

g) $(x^3 - x^2y + 5xy^2 + y^3) + (-x^3 - 9xy^2 + y^3) + (3x^2y + 9xy^2)$

h) $\left(\frac{2}{3}x^2 + xy + \frac{3}{2}y^2\right) + \left(\frac{5}{6}x^2 - \frac{3}{2}xy + 2y^2\right) + \left(4xy - 2x^2 - \frac{1}{2}y^2\right)$

i) $\left(\frac{5}{6} - \frac{7}{9}x^2 + ab\right) + \left(3ab + 2 - \frac{5}{9}x^2\right) + \left(5 - \frac{2}{3}x^2\right)$

**2** Simplify each of these calculations as much as you can.

a) $3xy^2 - 4xy^2$

b) $8x^2y - (-2x^2y + 3xy^2)$

c) $(7a - 3b + c - 2d) - (3a - 5b + c + d)$

d) $(4a + 6b - 2c) - (a - b - 2c)$

e) $(3x^3 - x^2 + 6) - (x^3 - 4x - 1)$

f) $(a^3 - 3a^2 + 4a + 1) - (6a + 3)$

g) $(3abc + 5bcd - cda) - (cab - 4cad - cdb)$

h) $(4a^2 - 3ab + 2b^2) - (a^2 + ab + b^2)$

i) $(-4x^2 + 5x - 6) - (4x^2 - 5x + 6)$

j) $(6a + 8b + 9c) - [(2b - 3a - 6c) - (c - b + 3a)]$

k) $(m^2 + m + 1) - [(-m^2 + 3m + 6) + (m^2 + m + 4)]$

l) $(6y^2 + y - 2) - [(5y^2 + y - 3) + (y^2 - 3y + 7)]$

m) $(ab - 4bc + 7cd) - [(6ab + 3bc - cd) - (15ab - bc + 8cd)]$

## H  Multiplying and dividing terms

To multiply a **constant term** (just a number) by a **variable term** (numbers and letters), multiply the number parts of each term as normal, and keep the letter part the same to make a new term.

**Examples**  Write each of these expressions in a simpler form.

a) $4 \times b$

$4 \times b = 4b$

b) $5 \times 2a$

$5 \times 2a = (5 \times 2) \times a$

$= 10a$

c) $(-3) \times 7y$

$(-3) \times 7y$

$= [(-3) \times 7] \times y$

$= -21y$

d) $(-2) \times (-8x^2)$

$(-2) \times (-8x^2)$

$= [(-2) \times (-8)] \times x^2$

$= 16x^2$

To multiply **two variable terms** together, multiply the two number parts (coefficients) together as normal, and then multiply the two variable letters together. If the letters are the same, use the correct power numbers to show how many of this letter are multiplied together.

**Examples**  Simplify each of these expressions.

a) $3x \times y$

$3x \times y = 3xy$

b) $5c \times 4c$

$$5c \times 4c = (5 \times 4) \times (c \times c)$$
$$= 20c^2$$

c) $(-a) \times b$

$$(-a) \times b = -ab$$

d) $(-b) \times (-b) \times b$

$$(-b) \times (-b) \times b = [(-1) \times (-1) \times (+1)] \times (b \times b \times b)$$
$$= +b^3$$

To divide a **variable term** by a **constant term**, divide the number parts as normal, and leave the variable.

**Examples**

Simplify each of these expressions.

a) $8a \div 2$

$$8a \div 2 = (8 \div 2) \times a$$
$$= 4a$$

b) $15p \div (-5)$

$$15p \div (-5) = 15 \div (-5) \times p$$
$$= -3p$$

c) $(-9x) \div 3$

$$(-9x) \div 3 = [(-9) \div 3] \times x$$
$$= -3x$$

d) $(-12y^2) \div (-4)$

$$(-12y^2) \div (-4) = [(-12) \div (-4)] \times y^2$$
$$= +3y^2$$

To divide **two variable terms**, divide the coefficients of each term as normal, and then use the power number rules to divide any powers of the same letter.

**Examples**

Simplify each of these expressions.

a) $9x \div x$

$$9x \div x = 9 \times (x \div x)$$
$$= 9$$

b) $6z \div (-2z)$

$$6z \div (-2z) = [6 \div (-2)] \times (z \div z)$$
$$= -3$$

c) $(-8a^2) \div 2a$

$$(-8a^2) \div 2a = [(-8) \div 2] \times (a^2 \div a)$$
$$= -4a$$

d) $(-15d^4) \div (-3d^2)$

$$(-15d^4) \div (-3d^2) = [(-15) \div (-3)] \times (d^4 \div d^2)$$
$$= +5d^2$$

**Exercise 6**

**1** Write each of these expressions in a simpler form.
a) $3 \times a$
b) $7 \times b$
c) $2 \times 4 \times c$
d) $3 \times 3 \times d$
e) $e \times 4$
f) $f \times 8$
g) $3 \times 2p$
h) $3q \times 5$
i) $r \times r$
j) $g \times g$
k) $2g \times g$
l) $3g \times 4g$
m) $t \times 4t$
n) $5u \times 3u$
o) $3d \times 3d$

**2** Simplify each of these expressions.
a) $3 \times (-y)$
b) $y \times (-5)$
c) $(-2) \times (-y)$
d) $3 \times (-2y)$
e) $t \times (-t)$
f) $2t \times (-t)$
g) $(-2t) \times 5t$
h) $(-2t) \times (-5t)$
i) $2 \times m \times (-5m^2)$

**3** Simplify each of these expressions.
a) $10a \div 2$
b) $16b \div 4$
c) $12x \div x$
d) $20y \div y$
e) $8y \div 4$
f) $8y \div 4y$
g) $18p \div p$
h) $18p \div 6$
i) $18k \div 2k$
j) $18a \div 3a$
k) $28g \div 7g$
l) $10m \div 2m$
m) $20t \div 5t$
n) $27x \div 3x$
o) $3d^2 \div 9d$

**4** Simplify each of these expressions.
a) $6y \div (-3)$
b) $(-6y) \div 2$
c) $(-5m) \div m$
d) $(-5m) \div (-5)$
e) $3a \div (-a)$
f) $(-10d) \div 5d$
g) $6g \div (-2g)$
h) $(-3k) \div (-6k)$
i) $12t^3 \div (-3t)$

**5** Simplify each of these expressions.
a) $a \times b$
b) $x \times y$
c) $y \times y \times 3$
d) $2 \times p \times q$
e) $2 \times a \times a$
f) $5 \times x \times y$
g) $3 \times a \times 2 \times b$
h) $3 \times g \times 4 \times b$
i) $2 \times d \times 3 \times d$
j) $3g \times g$
k) $a \times 5b$
l) $2g \times 3b$
m) $a \times b \times c$
n) $m \times m \times m$

o) $2 \times d \times d \times d$

p) $2x \times 3x \times x$

q) $g \times g \times g \times 3$

r) $5m \times m \times 2n$

s) $3a \times b \times c \times b$

t) $2p \times 3q \times 3r$

u) $7s \times 3t \times u \times t$

**6** Simplify each of these expressions.

a) $y^2 \times y$

b) $t^3 \times t^2$

c) $a^3 \times (-a^3)$

d) $g^7 \times g^3$

e) $a \times (-a) \times a^2$

f) $(-m) \times m^3 \times m^2$

g) $2y \times y^2$

h) $3d^2 \times 2d^3$

i) $4x^2 \times 2x^3$

j) $5y^3 \times (-5y^3)$

k) $2r \times (-3r^2) \times 4r^3$

l) $mn \times m^2n$

m) $a^2b \times ba^2$

n) $3rs^3 \times 2r^2s$

o) $2x^2y \times 5x^3y^2$

**7** Simplify each of these expressions.

a) $y^3 \div y$

b) $a^4 \div a^3$

c) $x^5 \div (-x^5)$

d) $t^7 \div t^3$

e) $g \div g^2$

f) $b^3 \div b^5$

g) $6b^3 \div b$

h) $10m^3 \div 2m^2$

i) $8x^8 \div 2x^2$

j) $16t^3 \div (-4t^2)$

k) $9b^2 \div 3b^3$

l) $xy^2 \div xy$

m) $m^3n \div mn^2$

n) $2p^3q \div 3pq^2$

o) $6r^2s^3 \div 2rs$

**8** Simplify each of these expressions.

a) $\dfrac{t^3}{t^2}$

b) $\dfrac{g^2}{g^3}$

c) $\dfrac{m^2 \times m}{m}$

d) $\dfrac{y^2 \times y^3}{y^4}$

e) $\dfrac{y \times y^3}{y^2}$

f) $\dfrac{m^2 \times m^3}{m^6}$

g) $\dfrac{2t^3 \times t}{t^2}$

h) $\dfrac{6g^2 \times (-gb)}{2g^3b^2}$

i) $\dfrac{3a^2b \times 4ab^3}{-2a^2b^2}$

## 1 Removing brackets using the distributive law

When a number is multiplied by a bracket, we can work out the answer in two ways.

● We can work out the contents of the bracket and then multiply.

For example, $2(5 + 3) = 2 \times 8$
$$= 16$$

● Or we can multiply the number outside the bracket by each of the numbers inside and then add the results.

For example, $2(5 + 3) = 2 \times 5 + 2 \times 3$
$$= 10 + 6$$
$$= 16$$

We call this second method **expansion**. It uses the distributive law.

We can use the distributive law in exactly the same way to remove the brackets from an expression in algebra.

For example, $2(3a + 4b) = 2 \times 3a + 2 \times 4b$
$$= 6a + 8b$$

Remember that the rules for multiplying in algebra are the same as the rules you have learned for multiplying numbers.

● positive × positive = positive    $(+2) \times (+3) = (+6)$
$(+a) \times (+b) = (+ab)$
● negative × negative = positive    $(-2) \times (-3) = (+6)$
$(-a) \times (-b) = (+ab)$
● positive × negative = negative    $(+2) \times (-3) = (-6)$
$(+a) \times (-b) = (-ab)$
● negative × positive = negative    $(-2) \times (+3) = (-6)$
$(-a) \times (+b) = (-ab)$

In general, the **distributive law** can be written as $a(b + c) = ab + ac$.

**Examples**    Expand each of these expressions.
a) $2(x + 1)$

$2(x + 1) = 2 \times x + 2 \times 1$
$$= 2x + 2$$

b) $4(3x - 5)$

$4(3x - 5) = 4 \times 3x + 4 \times (-5)$
$$= 12x - 20$$

If the number in front of a bracket is negative, take care with the signs when you multiply each term inside the bracket.

**Examples**

Expand each of these expressions.
a) $-2(a + 5)$
$$-2(a + 5) = (-2 \times a) + (-2 \times 5)$$
$$= -2a + (-10)$$
$$= -2a - 10$$
b) $-3(2b - 7)$
$$-3(2b - 7) = (-3 \times 2b) + (-3 \times -7)$$
$$= -6b + 21$$

Notice that the sign of each term in the answer is the opposite of the matching term inside the brackets in the question. The negative sign has made them change.

Sometimes there will be a **variable term** outside the brackets, as well as inside.

**Examples**

Expand each of these expressions.
a) $a(a + 3)$
$$a(a + 3) = (a \times a) + (a \times 3)$$
$$= a^2 + 3a$$
b) $b(2b - 3)$
$$b(2b - 3) = (b \times 2b) + (b \times -3)$$
$$= 2b^2 - 3b$$
c) $-c(c + 4)$
$$-c(c + 4) = (-c \times c) + (-c \times 4)$$
$$= -c^2 - 4c$$
d) $(2d + 1)d$
$$(2d + 1)d = (2d \times d) + (1 \times d)$$
$$= 2d^2 + d$$

Sometimes we need to expand more than one set of brackets, and then collect like terms.

**Examples**

Expand and simplify each of these expressions.
a) $4(2a - 3) + 3(a + 2)$
$$4(2a - 3) + 3(a + 2) = (4 \times 2a) + (4 \times -3) + (3 \times a) + (3 \times 2)$$
$$= 8a - 12 + 3a + 6$$
$$= 11a - 6$$

b) $2(a + 3b) - 3(b - a)$

$2(a + 3b) - 3(b - a) = 2a + 6b - 3b + 3a$
$= 5a + 3b$

c) $8x - 3(a + 5x)$

$8x - 3(a + 5x) = 8x - 3a - 15x$
$= -7x - 3a$

d) $b(b - 3) - b(2b + 5)$

$b(b - 3) - b(2b + 5) = b^2 - 3b - 2b^2 - 5b$
$= -b^2 - 8b$

## Exercise 7

**1** Expand each of these expressions by multiplying out the brackets.

a) $b(b + 7)$
b) $d(d - 4)$
c) $f(3f + 4)$
d) $3b(4b - 5)$
e) $-z(2z + 3)$
f) $-y(4y + 1)$
g) $-4k(3k - 5s)$
h) $-2a(b - 3c)$
i) $(4n + 3)n$
j) $5(x + y - 4)$
k) $2d(3d + 4e - 6de)$
l) $-3p(2m + 6p)$

**2** Expand and simplify each of these expressions.

a) $2(x + 1) + 3$
b) $3(a + 2) + 5$
c) $6(w - 4) + 7$
d) $4 + 2(p + 3)$
e) $3 + 3(q - 1)$
f) $1 + 3(2 - t)$
g) $4(z + 2) + z$
h) $5(t + 3) + 3t$
i) $3(c - 2) - c$
j) $2a + 3(a - 3)$
k) $y + 2(y - 5)$
l) $5x + 3(2 - x)$
m) $4(2a + 5) + 3$
n) $-2x + 4(3x - 3)$
o) $3(p - 5) - p + 4$
p) $3a + 2(a + b)$
q) $3(x + y) - 2y$
r) $2x + x(3 - x)$
s) $a(a - 3) + a$
t) $y(2 - y) + y^2$
u) $5d - 3(d - 2)$
v) $x^2 - x(1 - x) + x$
w) $t^2 - 2t(3 - 3t)$
x) $2m - 2m(m - 3)$

**3** Expand and simplify each of these expressions.

a) $5(x + 4) + 2(x + 7)$

b) $3(t - 5) + 8(t - 1)$

c) $4(r + 4) + 5(2r + 3)$

d) $5(2y - 5) + 8(3y - 1)$

e) $3(a + 2) - 2(a + 1)$

f) $4(a + 5) - 3(a + 1)$

g) $4(x + 3) - 2(x + 2)$

h) $11(t + 1) - 2(t + 5)$

i) $3m(2m + 1) - 4m(m - 1)$

j) $6n(3n + 5) - 3n(2n - 5)$

k) $3p(3p - 5) - 7p(2p + 3)$

l) $4y(y - 6) - 5y(3y - 1)$

m) $5a(2 - a) - 8a(a + 4)$

n) $3a(4 - 5a) - 2a(5 - 4a)$

o) $8z(2 - z) - 5z(3 - 2z)$

p) $2(3a + 1) + 3(a + 1)$

q) $3(2t + 5) + 5(4t + 3)$

r) $3(z + 5) - 2(z - 1)$

s) $7(q - 2) - 5(3q + 6)$

t) $5(x - 3) - 6(4x + 3)$

u) $8e(2e - 1) + 4(e - 2)$

v) $2(5d + 4) - 2d(d - 5)$

w) $m(m - 2) + 3m(2m - 1)$

x) $a(3a + 2) - 2a(7a - 3)$

# Unit 10 An introduction to geometry

People in ancient Greece were the first to start studying what we now call 'geometry'.

The word 'geometry' comes from two Greek words: 'geo' means 'earth', and 'metreo' means 'to measure'. So, geometry really means 'to measure the earth'!

The rules of geometry have helped people to do many things – finding our way in space or on the sea would be very difficult without geometry! Our buildings would not be so big or as safe without geometry either.

# Points, lines and planes

All geometry is made up of three things: points, lines and planes.

## Points

- A **point** has **no** length, **no** width and **no** depth (thickness).
- Even though it has no size, each point shows us one special 'place'.
- We use a dot to show where a point is (although the real point is much smaller than this dot), and give it a name using a capital letter (e.g. *S*).

## Lines

- A **line** has **length**, but has **no width** and **no thickness**.
- We can think of a line as a lot of points all joined up together in a row.
- A line goes on forever in both directions. To show this, we draw arrows on both ends. We use a lower case letter to name a line (e.g. *l*).

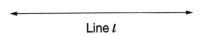

Line *l*

- Sometimes we look at only **part** of a line, starting at one **point** and going on forever in **one direction only**. We call this a ray. We use the starting point and one other point on the ray to give it a name. Usually, we draw an arrow above the names of these two points to show that it is a ray.

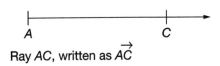

Ray *AC*, written as $\overrightarrow{AC}$

- A **line segment** has a definite length and is defined by one point at each end of the segment.

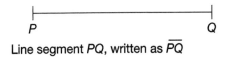

Line segment *PQ*, written as $\overline{PQ}$

**Planes**
- A **plane** has both **width** and **length**, but no **thickness**.
- A plane is like a very big flat surface. We can think of it as a very big piece of paper that goes on forever in all directions.
- Planes can lie in any direction, but the most common planes are horizontal and vertical.
- The most important planes are the *x*-plane, the *y*-plane and the *z*-plane. We use these planes to find any point in 3-dimensional space.

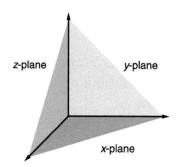

## Measuring the length of a line segment

I am sure you have measured **line segments** many times before (you probably called them 'lines' for short – but a **line** continues forever in both directions, so we can't measure it!)

### Using only a ruler
This is the way that most people measure a line segment.
1 Place the 0 mark of a ruler on one end of the line segment.
2 Then read the length on the ruler at the other end of the line segment.

### Using a pair of dividers (or compasses) and a ruler
This is the way we will measure a line segment in geometry class.
It is best to use a pair of dividers for measuring if you have them
– if not, use a pair of compasses instead.
1 Place one point of the dividers (or compasses) on one end of the line segment.
2 Open the dividers to place the other point exactly on the other end of the line segment.
3 Place one point of the dividers carefully on the 0 mark of a ruler.
4 Read the length on the ruler at the other point of the dividers.

**Exercise 1**

1 Carefully measure the length of each of these line segments.
Remember to include the **units** in your answer.

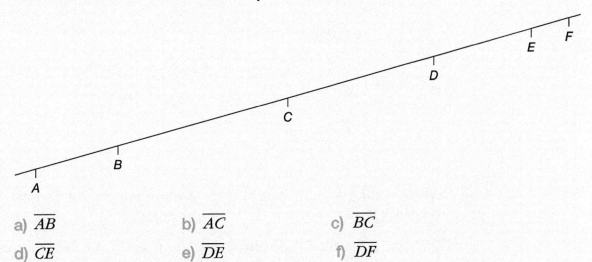

a) $\overline{AB}$          b) $\overline{AC}$          c) $\overline{BC}$

d) $\overline{CE}$          e) $\overline{DE}$          f) $\overline{DF}$

2 Use a pair of dividers (or compasses) and a ruler to measure the length of
each of these line segments.

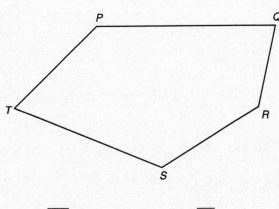

a) $\overline{PQ}$          b) $\overline{QR}$          c) $\overline{RS}$

d) $\overline{ST}$          e) $\overline{TP}$

3 Write down all the line segments that you can find in this diagram.

4 Use your answer to question 3 to help you work out the number of line
segments you could find if there were 20 points on the line.
HINT: Start with 2 points, 3 points, 4 points, 5 points and so on until you can
see the pattern.

## B Angles

If two line segments or rays start or meet at one point, an **angle** will be formed.

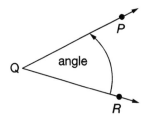

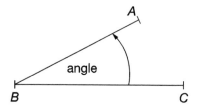

Rays $\overrightarrow{QP}$ and $\overrightarrow{QR}$ both start at the point $Q$.

The point $Q$ is called the **vertex**.

Line segments $\overline{BA}$ and $\overline{BC}$ meet at point $B$.

The point $B$ is called the **vertex**.

The **symbols** we use to write the name of an angle are: $\angle$ or $\hat{\phantom{x}}$.

This angle is written as $\angle PQR$ or $P\hat{Q}R$.

This angle is written as $\angle ABC$ or $A\hat{B}C$.

We can also use a short way to write the names of angles.

$\angle PQR$ can also be written as $\angle Q$.

$P\hat{Q}R$ can also be written as $\hat{Q}$.

$\angle ABC$ can also be written as $\angle B$.

$A\hat{B}C$ can also be written as $\hat{B}$.

## C Types of angles

We can explore the different types of angles using a homemade tool.

- Cut two pieces of cardboard. One piece should be 2 cm wide and 20 cm long. The other piece should be 2 cm wide and 15 cm long.
- Join them together with a paper fastener like this.

Keep the long strip still, and move the short strip a small way apart.

This will make a very 'sharp' or small angle.

A small angle like this is called an acute angle.

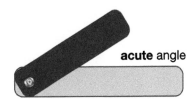

**acute** angle

Now move the short strip a little further so that it is standing straight up and the other one is lying down.

This is a very 'square' angle.

A square angle like this is called a right angle.

If one line is at right angles to another, they are perpendicular to each other.

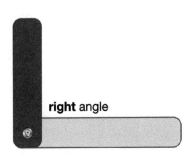

**right** angle

Now move this strip a little further away so that it is leaning away from the other one.

This makes a bigger and wider angle.

This is called an obtuse angle.

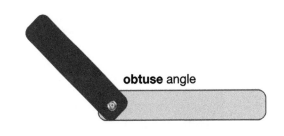

**obtuse** angle

We find our next type of angle when we move both strips so that they are in a straight line.

This angle is called a straight angle.

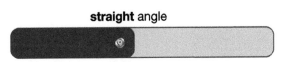

**straight** angle

If we move the one strip a little further, it will look as if it is falling down.

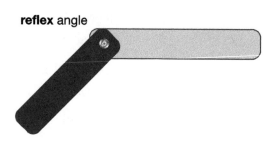

**reflex** angle

This is a very big angle (but you will see that there is a smaller angle on the other side).

This big angle is called a reflex angle.

Now, if we move the strips so that they are back where they started, they have made a **full turn**.

**full turn**

**Exercise 2**

1 Use letters to name each of these angles in two different ways.

a)
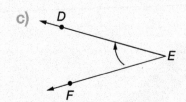

b)
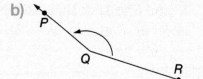

c)

d)
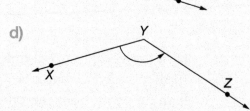

2 Write down what type of angle each of these is (acute, obtuse, reflex or a right angle).

a)

b)
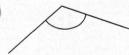

c)

d)

e)

f)

g)

h)

## D Measuring angles

An angle is made by **turning** one ray further and further away from another ray while the **vertex point** stays in the same place.
This circular movement is called rotation.

The biggest angle we can make is when a ray rotates all the way around and back to where it started. We call this a **full turn**.
It is the same as a full circle.

This means that every angle is a **part of a circle**.

full circle          angle
part of a circle

Many thousands of years ago, ancient mathematicians decided to divide a circle into **360** equal parts (like slices). Each of these parts is called a degree. Because an angle is part of a circle, we can use these 360 degrees (written as 360°) in a circle to measure how big an angle is.

A straight angle is half a turn, or half a circle. So a **straight angle** is:
$$\frac{1}{2} \times 360° = \mathbf{180°}$$

A right angle is a quarter turn, or quarter circle. So a **right angle** is:
$$\frac{1}{4} \times 360° = \mathbf{90°}$$

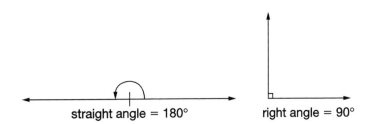

straight angle = 180°          right angle = 90°

We measure the size of an angle with a protractor.

If you look at a protractor, you will see that there are two sets of numbers – one on the inside and one on the outside. Each set of numbers is called a scale and can be used to measure angles from 0° to 180°. We use the outside scale to measure angles that are open to the left, and the inside scale to measure angles that are open to the right.

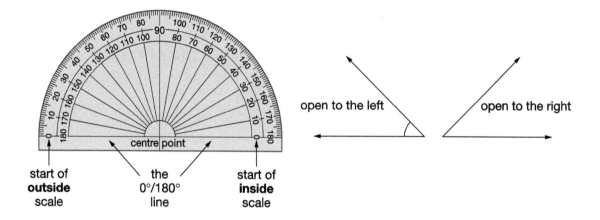

start of **outside** scale

the 0°/180° line

start of **inside** scale

open to the left

open to the right

To measure an angle, it is important to place the protractor on the lines of the angle in exactly the right place.

- One **line** of the angle must be exactly along the **0°/180° line** of the protractor.
- The **vertex** of the angle must be exactly on the **centre point** of the protractor.

Take care to use the correct scale (inside or outside). Check that your measured answer agrees with these values for each type of angle.
- If the angle is an **acute** angle, it will be < 90°.
- If the angle is an **obtuse** angle, it will be > 90° but < 180°.

If you are measuring an **acute** angle, and you read it as 135°, you know that you have read the wrong scale. The answer should be 45° (on the other scale).

When we measure an angle that opens to the left, we always read the **outside** scale on the protractor.

The angle below is 52°, **not** 128°.

Check: the angle is an **acute** angle, and so it must be < 90°.

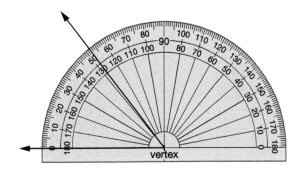

This angle also opens to the left. Again, we will use the **outside** scale on the protractor.

The angle below measures 125°, **not** 55°.

Check: the angle is an **obtuse** angle, and so it must be > 90°.

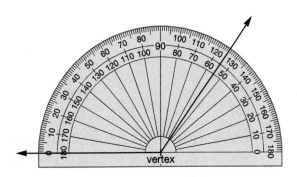

When we measure an angle that opens to the right, we always read the **inside** scale on the protractor.

The angle below is 43°, **not** 137°.

Check: the angle is an **acute** angle, and so it must be < 90°.

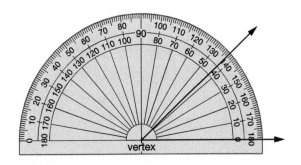

This angle also opens to the right. Again, we will use the **inside** scale on the protractor.

The angle below measures 155°, **not** 25°.

Check: the angle is an **obtuse** angle, and so it must be > 90°.

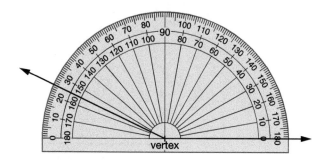

## Measuring a reflex angle

The scale of most protractors only goes up to 180°. Reflex angles are bigger than 180° – so we have to measure them in a slightly different way.

Look at this reflex angle.

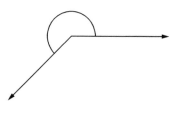

Extend one of the lines of the angle beyond the **vertex** to make a straight angle (180°).

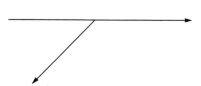

Now measure the smaller angle as normal with your protractor.

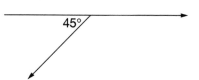

The full angle is 180° + 45° = 225°.

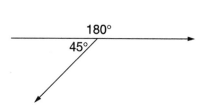

If the reflex angle is bigger than 270°, it is easier to measure the **smaller** angle made by the two lines.

Look at this reflex angle.

A full turn is 360°, so the bigger angle (the reflex angle) is 360° − 45° = 315°.

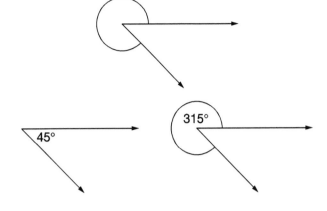

## Major and minor angles

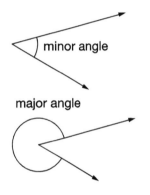

Every time two rays meet at a vertex point, **two** angles are formed – a **smaller** one and a **bigger** one.

We call the smaller angle the 'minor angle' and we call the bigger angle the 'major angle', as shown on the left.

## E  Drawing angles

We can also use a protractor to **draw** angles.

**Example**

Draw an angle of 72°.

1 First, draw a line segment. Mark a point that will be the vertex of the angle.

2 Place the 0°/180° line of the protractor on your line segment, with the vertex point at the centre. Mark a dot at 72°.

3 Draw a line to join the vertex point and the dot at 72°. You have drawn an angle of 72°.

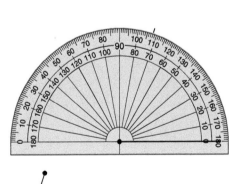

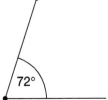

## Exercise 3

**1** Use a protractor to measure the size of these angles.

a) ∠*DOC*

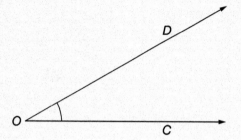

b) *Ô*

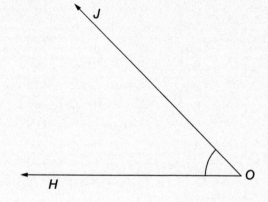

c) *AB̂C*

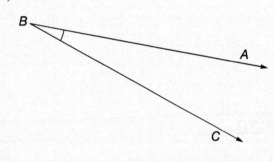

d) ∠*XYZ*

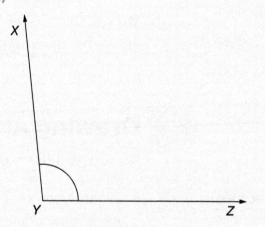

e) ∠*PQR*

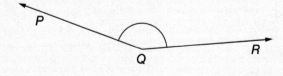

f) ∠*TNK*

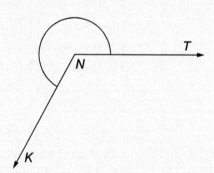

g) *FÊR*

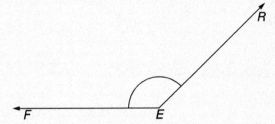

h) ∠*U*

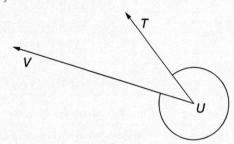

**2** Use a protractor to draw and label each of these angles.

a) ∠*ABC* = 36°           b) ∠*XYZ* = 85°

c) ∠*CDE* = 123°          d) ∠*PQR* = 292°

## F Some special angles

Sometimes angles have special names because they are in special places together with other angles.

NOTE: When we add the sizes of two or more angles, we call this the **sum** of the angles.

### Adjacent angles

If the **same ray** is part of two different angles, and the **vertex** of both angles is the same point, the two angles are called adjacent angles.

∠*XAY* and ∠*YAZ* are adjacent angles.

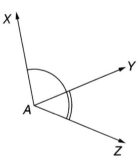

### Complementary angles

If the sum of two or more adjacent angles is 90° exactly (a right angle), these angles are called complementary angles. When there are only two angles, each angle is called the complement of the other.

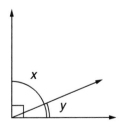

∠*x* + ∠*y* = 90°, so ∠*x* and ∠*y* are complementary angles.
∠*x* is the complement of ∠*y*, and ∠*y* is the complement of ∠*x*.

$\angle a + \angle b + \angle c = 90°$, so
$\angle a, \angle b$ and $\angle c$ are
complementary angles.

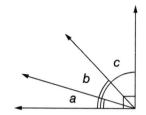

## Supplementary angles

If the sum of two or more
adjacent angles is 180° exactly
(a straight line), these angles are
called supplementary angles.
When there are only two angles,
each angle is called the
supplement of the other.

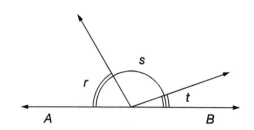

$\overline{AB}$ is a straight line, so $\angle r + \angle s + \angle t = 180°$.

So $\angle r, \angle s$ and $\angle t$ are supplementary angles.

## Angles at a point

When adjacent angles meet at **one
point**, the sum of the angles is always
360°.

$\angle u, \angle v$ and $\angle w$ all meet at one point.
So we know that
$\angle u + \angle v + \angle w = 360°$.

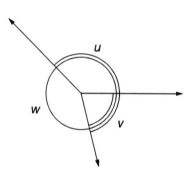

## Vertically opposite angles

When two straight line segments cross at
a point, it makes **two pairs** of **equal
angles** that are directly opposite each
other. We call these angles vertically
opposite angles.

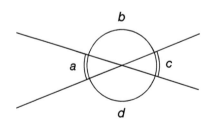

$\angle a = \angle c$ and $\angle b = \angle d$

$\angle a$ and $\angle c$ are vertically opposite angles.
$\angle b$ and $\angle d$ are vertically opposite angles.

NOTE: We know that angles are made when two rays or two line segments meet at a point. However, it is not always necessary to show the arrows for the rays, or the marks at the ends of the line segments when we draw angles. We often draw angles more simply, like this.

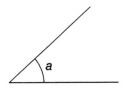

We can use these **angles facts** to find the size of unknown angles.

**Examples**   **a)** Find the angle marked '*a*' in each of these diagrams.

i)

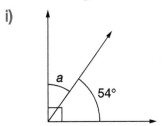

ii)

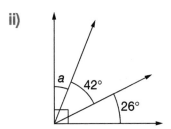

i)   $\angle a + 54° = 90°$ (complementary)
So $\angle a = 90° - 54°$
$= 36°$

ii)   $\angle a + 42° + 26° = 90°$ (complementary)

So $\angle a = 90° - (42° + 26°)$
$= 90 - 68°$
$= 22°$

**b)** Find the complement of an angle that measures 17°.

The complement of an angle is another angle such that the sum of their sizes is 90°.
The complement of $17° = 90° - 17° = 73°$

c) Find all the angles marked in each of these drawings.

i)

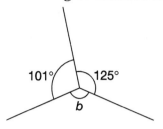

ii)

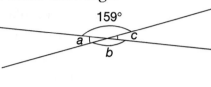

i)    Sum of angles at a point = 360°
$$\angle b + 101° + 125° = 360°$$
$$\text{Or } \angle b = 360° - (101° + 125°)$$
$$= 360° - 226°$$
$$= 134°$$

ii)
$$\angle b = 159° \text{ (vertically opposite)}$$
$$\angle a + 159° = 180° \text{ (supplementary angles)}$$
$$\text{So } \angle a = 180° - 159°$$
$$= 21°$$
$$\angle c = \angle a \text{ (vertically opposite)}$$
$$= 21°$$

d) Find the supplement of an angle that measures 84°.

The supplement of an angle is another angle such that the sum of their sizes is 180°.
The supplement of 84° = 180° − 84°
$$= 96°$$

## Exercise 4

1 Find the complement of each of these angles.
  a) 10°        b) 60°        c) 65°        d) 12°
  e) 74°        f) 83°        g) 5°        h) 38°

2 Find the supplement of each of these angles.
  a) 12°        b) 26°        c) 118°        d) 104°
  e) 136°      f) 153°      g) 179°      h) 61°

3 Write down whether each of these pairs of angles is complementary or supplementary.
  a) 120°, 60°    b) 70°, 110°    c) 30°, 60°    d) 80°, 10°
  e) 142°, 38°    f) 93°, 87°    g) 78°, 102°    h) 81°, 99°
  i) 49°, 41°    j) 52°, 38°

4 Without measuring, work out the size of each angle marked with a letter.

a)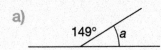

149° a

b)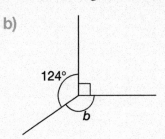

124°

b

c)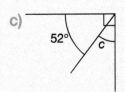

52°

c

d)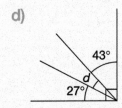

43°

d

27°

e)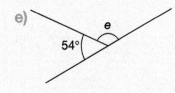

54°

e

f)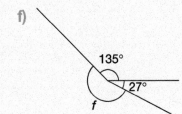

135°

27°

f

g)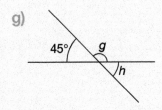

45°

g

h

h)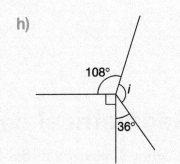

108°

i

36°

i)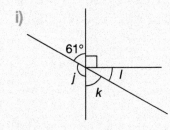

61°

j

k

l

j)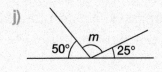

50°

m

25°

k)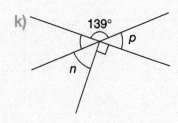

139°

p

n

l)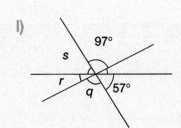

97°

s

r

q

57°

**5** Without measuring, work out the value of $x$ in each diagram.

a)
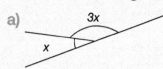

b)

c)
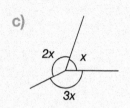

d)
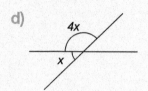

e)
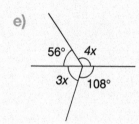

f)
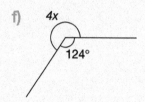

g)
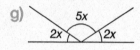

h)
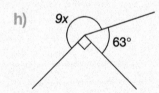

**G** # Geometrical constructions

Sometimes we need to draw special lines and angles **accurately**.
To do this, we use only a **sharp pencil**, a **ruler** and a **pair of
compasses**. An accurate diagram drawn in this way is called
a construction.

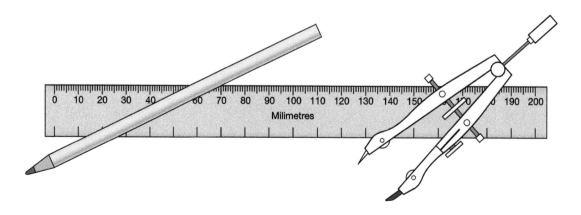

## To copy an angle

We have already learned how to draw an angle using a protractor.

If we are given an angle already drawn, it is possible to make a copy of it using just a ruler and a pair of compasses. This is more accurate than using a protractor.

We are going to copy $\angle X$.

1 Place the point of your compasses on the vertex of the angle ($X$). Draw an arc. Label the points at which your arc crosses the rays of the angle points $Y$ and $Z$.

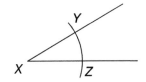

2 We will call our copy $\angle A$. Draw a new line segment $\overline{AL}$. Place the point of your compasses on the point $A$ (it will be the vertex of the new angle) and draw an arc exactly the same size as the one you drew on the first angle. This arc cuts the line $\overline{AL}$ at the new point $B$.

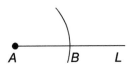

3 Set the compasses so that they are exactly the same length as $ZY$. Now place the point of your compasses on point $B$ and draw an arc that crosses the first arc at the new point $C$.

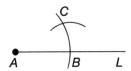

4 Draw a line through points $A$ and $C$. You can use a protractor to check that $\angle A = \angle X$.

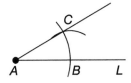

## To draw the bisector of an angle

The bisector of an angle is a line drawn through the vertex that divides the angle into **two equal parts**. We are going to bisect angle $A$.

**1** Place the sharp point of your compasses on the point $A$. Use the compasses to mark two points $B$ and $C$. These points can be anywhere along the rays that make the angle, but they must be the same distance from point $A$.

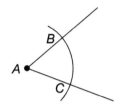

**2** Now draw equal arcs from points $B$ and $C$. The arcs cross at point $D$.

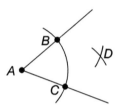

**3** Draw a line that passes through points $A$ and $D$. This line divides $\angle BAC$ exactly into two equal parts. So $\angle BAD = \angle DAC$ (use a protractor to check). $\overline{AD}$ is the **bisector** of $\angle BAC$.

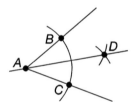

## To draw the perpendicular bisector of a line segment

A line that is drawn at right angles to a given line segment, and divides it exactly into two equal parts, is called the perpendicular bisector of that line segment. We are going to bisect $\overline{AB}$.

**1** Open your compasses so that they are a little bit wider than half the distance $AB$. With the point of your compasses on point $A$, mark one arc above the line segment and one arc below it. Mark two arcs in the same way with the point of your compasses on point $B$. The two arcs above the line segment cross at $C$ and the two arcs below the line segment cross at $D$.

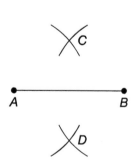

**2** Draw a line through the points $C$ and $D$. The line $CD$ is at right angles to $\overline{AB}$ and crosses it at exactly the middle point (mid point). Line $CD$ is the **perpendicular bisector** of $\overline{AB}$.

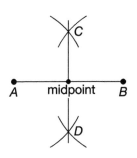

## To draw the perpendicular from a point to a line segment

This line will be at **right angles** to the given line segment, and passes through a given point that is **not** on the line segment.
We are going to draw the perpendicular to $\overline{PQ}$ that passes through point $A$.

**1** Place the point of your compasses on the given point $A$ and open them so that you can mark two points on $\overline{PQ}$.

**2** Now place the point of your compasses on the points $B$ and $C$ in order to draw two arcs that cross at point $D$.

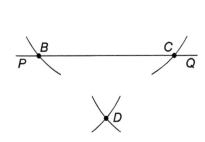

**3** Draw a line that passes through points $A$ and $D$. This line is at right angles to $\overline{PQ}$ (and of course it passes through the point $A$).

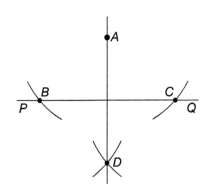

# To draw the perpendicular at a point on a line segment

This time we will draw a line at **right angles** to the given line segment, but now passing through a point **on** the line segment. We are going to draw the perpendicular to $\overline{PQ}$ that passes through point $A$.

**1** Place the point of your compasses on the given point $A$ and open them so that you can draw an arc that crosses the line segment $PQ$ at the point $B$.

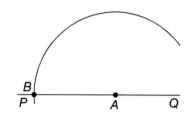

**2** Keep the compasses open exactly the same amount, and place their point on the point $B$. Draw another arc that crosses the first one at the point $C$. Now place the point of your compasses at the point $C$ and draw a third arc that crosses the first one at the point $D$.

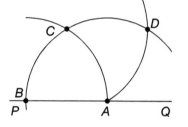

**3** From $C$ and $D$, draw arcs that cross at the point $E$. Draw a line that passes through points $A$ and $E$. This line is at right angles to $PQ$ (and of course it passes through the point $A$).

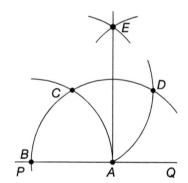

## To draw a triangle

To make an **accurate** drawing of a triangle, we must have some **measurements**.

**If we know the lengths of all three sides**
Draw $\triangle ABC$ with $AB = 5\,cm$, $BC = 4\,cm$ and $CA = 2\,cm$.

**1** First draw the line $AB$ exactly 5 cm long.
**2** Set your compasses to exactly 2 cm wide. Place their point on $A$ and draw an arc.
**3** Now set your compasses to exactly 4 cm wide. Place their point on $B$ and draw an arc so that it crosses the first arc to make point $C$.
**4** Join $A$ to $C$, and $B$ to $C$ to make the triangle $ABC$.

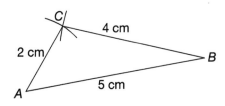

**If we know the lengths of two sides and the angle between them**
Draw $\triangle ABC$ with $AB = 3\,cm$, $AC = 2\,cm$ and $\angle BAC = 80°$.

**1** First draw the line $AB$ exactly 3 cm long.
**2** Use a protractor to draw an angle of exactly 80° at $A$.
**3** Set your compasses to exactly 2 cm wide. Place their point on $A$ and draw an arc to cross the ray of the angle at the point $C$.
**4** Join $B$ to $C$ to make the triangle $ABC$.

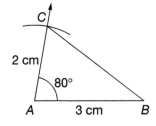

**If we know two angles and the length of the side between them**
Draw $\triangle ABC$ with $AB = 4\,cm$, $\angle CAB = 80°$ and $\angle CBA = 30°$.

**1** First draw the line $AB$ exactly 4 cm long.
**2** Use a protractor to draw an angle of exactly 80° at $A$.
**3** Use a protractor to draw an angle of exactly 30° at $B$.
**4** The rays of these two angles will cross at point $C$.
This will make the triangle $ABC$.

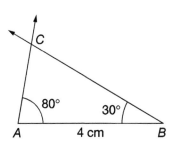

## To draw a square or a rectangle

Remember that a square is just a special rectangle with **all four sides** of **equal** length.

All the angles in a rectangle are right angles (90°), so we will use what we have learned about constructing **perpendicular lines** to construct an accurate drawing of the rectangle *ABCD*, with *AB* = 5 cm and *BC* = 4 cm.

**1** First draw the line *AB* exactly 5 cm long.
**2** At the point *B*, construct a perpendicular line. Now open you compasses to exactly 4 cm. Place their point on *B* and mark an arc on the perpendicular line so that it crosses at point *C*.
**3** Place the point of your compasses on *A* and make a similar arc directly above *A*.
**4** Now open your compasses to exactly 5 cm and place their point on *C*. Make an arc so that it crosses the one above *A* to give the point *D*.
**5** Join *A* to *D*, and *D* to *C* to make the rectangle.

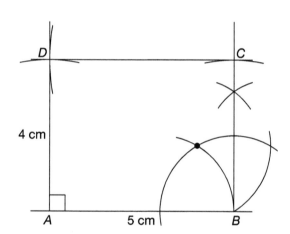

 **Exercise 5**

1 Use a protractor to draw an angle of 40°. Bisect the angle using only compasses and a ruler. Use your protractor to check that both halves of the angle are equal.

2 Use a protractor to draw an angle of 70°. Bisect the angle using only compasses and a ruler. Use your protractor to check that both halves of the angle are equal.

3 Draw a line segment 9 cm long. Bisect this line segment using only compasses and a ruler. Check that the lengths of the two halves are equal.

4 Draw a line segment 11 cm long. Bisect this line segment using only compasses and a ruler. Check that the lengths of the two halves are equal.

**5** Draw a line segment 10 cm long. Mark a point *A* about 6 cm above the line segment. Construct a perpendicular line to the line segment, from *A*.
Use only compasses and a ruler. Use your protractor to check that the line is perpendicular.

**6** Draw a line segment 12 cm long. Mark a point *A* about 6 cm above the line segment. Construct a perpendicular line to the line segment, from *A*.
Use only compasses and a ruler. Use your protractor to check that the line is perpendicular.

**7** Draw a line segment 12 cm long. Mark a point *A* on the line segment about 5 cm from the left-hand end. Construct a perpendicular line at the point *A* using only compasses and a ruler. Use your protractor to check that the line is perpendicular.

**8** Draw a line segment 10 cm long. Mark a point *A* on the line segment about 4 cm from the right-hand end. Construct a perpendicular line at the point *A* using only compasses and a ruler. Use your protractor to check that the line is perpendicular.

**9** Using compasses and a ruler only, make an accurate drawing of the triangle shown in this rough sketch.

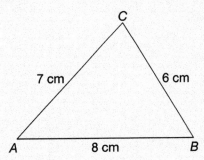

**10** Using compasses and a ruler only, make an accurate drawing of the triangle shown in this rough sketch.

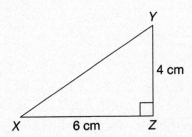

**11** Use compasses and a ruler to construct $\triangle ABC$ with $AB = 3\,\text{cm}$, $BC = 5\,\text{cm}$ and $AC = 7\,\text{cm}$.

**12** Use compasses and a ruler to construct $\triangle ABC$ with $AB = 4\,\text{cm}$, $AC = 6\,\text{cm}$ and $\angle BAC = 50°$.

**13** Use compasses and a ruler to construct $\triangle ABC$ with $AB = 6\,\text{cm}$, $\angle CAB = 40°$ and $\angle CBA = 60°$.

**14** Use compasses and a ruler to construct a rectangle $EFGH$ with $EF = 5.5\,\text{cm}$ and $FG = 3.3\,\text{cm}$.

**15 a)** Follow these steps to draw an angle $XPY$.
- Draw a line segment $PQ$, 5 cm long.
- Place the point of your compasses at point $P$.
- Draw an arc so that it crosses $PQ$ at $X$.
- With the compasses at the same width, place the point at $X$ and draw another arc so that it crosses the first one at $Y$.
- Draw a line through $PY$.
- Use a protractor to measure $\angle XPY$.

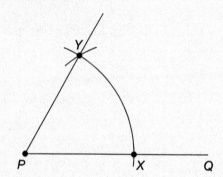

**b)** Think of a way to use this drawing to draw angles of 30° and 120°.

# An introduction to coordinate geometry

## Key vocabulary

| | | |
|---|---|---|
| axis | intersect | *x*-axis |
| Cartesian plane | ordered pair | *x*-coordinate |
| coordinates | origin (*O*) | *y*-axis |
| directed number | plot | *y*-coordinate |
| graph | quadrant | |
| horizontal | vertical | |

 **A grid of squares**

More than 400 years ago, a mathematician in France worked out a way to describe exactly where something is.

He was lying in bed looking at the ceiling. He saw a fly crawling on the ceiling and thought about how he could describe exactly where the fly was.

If he imagined a 'grid' of squares covering the ceiling, then he could count how many squares the fly was away from the walls.

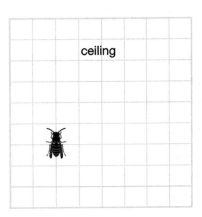

To count the squares, he had to decide where to start! There are four corners to the ceiling, and he could start counting from any of these four places.

So he made a **rule** – and we still use this rule today.

  ● Start at the **bottom left-hand** corner.
  ● Count the squares **across** first.
  ● Then count the squares **up**.

The fly in our picture is **2 squares to the right and 3 squares up**.

We can use grids like this to show exactly where things are, and how to **move** from one point to another.

##  Directed numbers and the Cartesian plane

A directed number is one that tells us **how far** to move, and what **direction** to move, from a point.

We met directed numbers when we learned about **positive** and **negative** numbers on the **number line** (see Unit 1).

Remember:
A **positive** (+) number is on the **right**-hand side of the zero point.
A **negative** (−) number is on the **left**-hand side of the zero point.

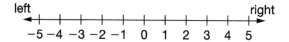

We can see that **right** and **left** are opposite directions. But **up** and **down** are also opposite directions. This means that we can use two number lines at right angles to each other that will help us to measure distance left and right, **and** up and down.

If they cross (intersect) at the zero point on each line, this will give a special point that is **no distance** left or right and **no distance** up or down. This point is called the origin (O) as this is where we start if we want to find any other point.

Each of these two perpendicular number lines is now called an axis.

This system with a horizontal and a vertical axis on a plane is called the Cartesian plane, after the French mathematician who first came up with the idea.

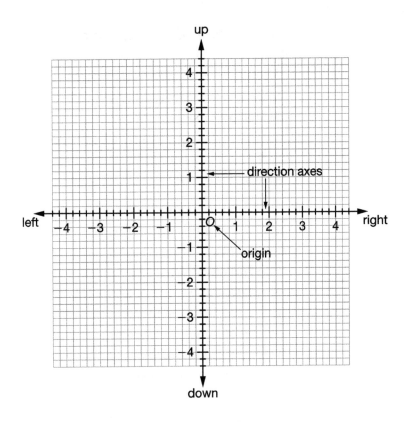

As we learned above, the **position** of a point on a plane can be described using two measurements. We can make these measurements using our two **axes**.

Look at the point $P$ marked on this diagram.

Point $P$ is 2 units to the **right** of the origin. This is the **positive** direction, so we call this $+2$.

Point $P$ is also 1 unit **up** from the origin. This is also in the **positive** direction, so we call this $+1$.

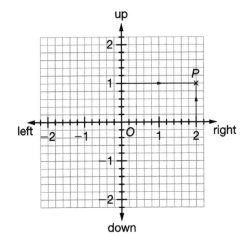

We write the position of the point $P$ by using **both** of these numbers.

The **right/left** instruction is always written **first**, and the **up/down** instruction is always written **second**. Both numbers are written in brackets. So the position of point $P$ is $(+2, +1)$.

The pair of numbers made in this way is called an ordered pair of directed numbers.

So the position of a point can be written using an **ordered pair** that tells us

- the horizontal (left/right) **distance** from the origin (using numbers 1, 2, 3, …, with fractions or decimals if necessary) and the **direction** (using + or −).
- the vertical (up/down) **distance** from the origin (using numbers 1, 2, 3, …, with fractions or decimals if necessary) and the **direction** (using + or −).

It would take a long time to describe the position of points if we had to say 'horizontal' and 'vertical' all the time, so instead we use the letters $x$ and $y$.

- The **horizontal** axis (left/right) is called the $x$-axis.
- The **vertical** axis (up/down) is called the $y$-axis.
- The **first** number in the ordered pair tells us about the movement on the $x$-**axis**. This number is called the $x$-coordinate of the point.
- The **second** number in the ordered pair tells us about the movement on the $y$-**axis**. This number is called the $y$-coordinate of the point.
- Together, this ordered pair $(x, y)$ is simply called the coordinates of the point.
- On the $x$-**axis**, any point to the **right** of the origin is **positive** (+) Any point to the **left** of the origin is **negative** (−).
- On the $y$-**axis**, any point **above** the origin is **positive** (+) Any point **below** the origin is **negative** (−).
- When we draw a point on a Cartesian plane using its coordinates, we say that we plot the point.
- When we plot many points and join them with lines, we call this a graph.

**Examples**

Write down the coordinates of each of the points marked on this Cartesian plane.

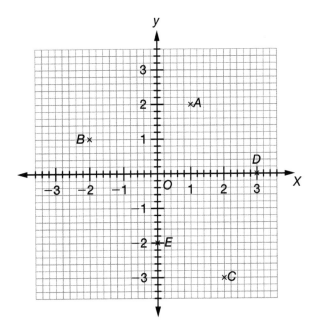

- Point *A* is 1 unit to the right of the origin, so the *x*-coordinate will be +1. It is 2 units above the origin, so the *y*-coordinate will be +2. So the coordinates of *A* are (+1, +2) or (1, 2).
- Point *B* is 2 units to the left of the origin, so the *x*-coordinate will be −2. It is 1 unit above the origin, so the *y*-coordinate will be +1. So the coordinates of *B* are (−2, +1) or (−2, 1).
- Point *C* is 2 units to the right of the origin, so the *x*-coordinate will be +2. It is 3 units below the origin, so the *y*-coordinate will be −3. So the coordinates of *C* are (+2, −3) or (2, −3).
- Point *D* is 3 units to the right of the origin, so the *x*-coordinate will be +3. It is 0 units above the origin (it is on the *x*-axis), so the *y*-coordinate will be 0. So the coordinates of *D* are (+3, 0) or (3, 0).
- Point *E* is 0 units from the origin (it is on the *y*-axis), so the *x*-coordinate will be 0. It is 2 units below the origin, so the *y*-coordinate will be −2. So the coordinates of *E* are (0, −2).

## C Quadrants

The *x*-axis and the *y*-axis divide the Cartesian plane into four pieces.

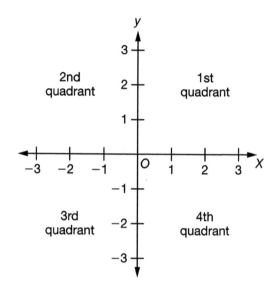

These four pieces are called quadrants.

Each quadrant is numbered, starting from the top right.

**1st quadrant**
- All *x*-coordinates in this quadrant are positive.
- All *y*-coordinates in this quadrant are positive.

**2nd quadrant**
- All *x*-coordinates in this quadrant are negative.
- All *y*-coordinates in this quadrant are positive.

**3rd quadrant**
- All *x*-coordinates in this quadrant are negative.
- All *y*-coordinates in this quadrant are negative.

**4th quadrant**
- All *x*-coordinates in this quadrant are positive.
- All *y*-coordinates in this quadrant are negative.

÷ %

### Exercise 1

**1** Write down the coordinates of each of the points *A* to *L* marked on this Cartesian plane.

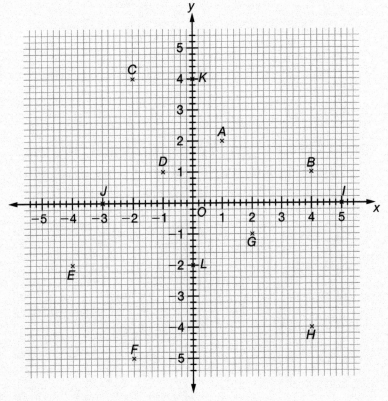

**2** Write down the coordinates of each of the points *A* to *L* marked on this Cartesian plane.

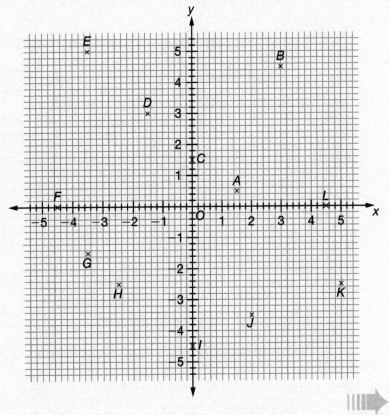

**3** On a piece of graph paper, draw and label the *x*- and *y*-axes to make a Cartesian plane. Now plot (and label) each of these points.

A(3, 1)                 B(4, 5)
C(2, 0)                 D(0, 0)
E(−4, 2)                F(−3, 1)
G(−2, 0)                H(−4, 0)
I(3, −2)                J(2, −5)
K(0, −4)                L(−3, −2)

**4** On a piece of graph paper, draw and label the *x*- and *y*-axes to make a Cartesian plane. Now plot (and label) each of these points.

$M\left(5, 3\frac{1}{2}\right)$        $N\left(4\frac{1}{2}, -2\right)$

$P\left(-1\frac{1}{2}, 3\right)$        $Q\left(-3, -4\frac{1}{2}\right)$

$R\left(0, 2\frac{1}{2}\right)$        $S\left(4\frac{1}{2}, 0\right)$

$T\left(1\frac{1}{2}, -4\frac{1}{2}\right)$        $U\left(-2\frac{1}{2}, 1\frac{1}{2}\right)$

$V\left(-3\frac{1}{2}, -1\frac{1}{2}\right)$        $W\left(-1\frac{1}{2}, 0\right)$

$X\left(-4, -1\frac{1}{2}\right)$        $Y\left(-3\frac{1}{2}, -2\right)$

**5** Copy and complete this table. The first two points have been completed for you.

| Point | x-coordinate (+ or −) | y-coordinate (+ or −) | Quadrant |
|---|---|---|---|
| A(3, 2) | + | + | 1st |
| B(−1, 4) | − | + | 2nd |
| C(2, −3) | | | |
| D(−4, −1) | | | |
| E(2, −5) | | | |
| F(−3, 1) | | | |
| G(5, −2) | | | |
| H(−6, −3) | | | |
| I(−2, 4) | | | |
| J(1, 5) | | | |
| K(4, −1) | | | |

**6 a)** On a piece of graph paper, draw and label the *x*- and *y*-axes to make a Cartesian plane. Now plot (and label) each of these points.

| | |
|---|---|
| *A*(0, 4) | *B*(2, 1) |
| *C*(5, 1) | *D*(3, −2) |
| *E*(4, −5) | *F*(0, −3) |
| *G*(−4, −5) | *H*(−3, −2) |
| *I*(−5, 1) | *J*(−2, 1) |

**b)** Join the points in alphabetical order, with straight lines. (Join *A* to *B*, *B* to *C*, *C* to *D*, … and so on. Finish by joining *J* to *A*.) What picture have you drawn?

**7 a)** On a piece of graph paper, draw and label the *x*- and *y*-axes to make a Cartesian plane. Now plot (and label) each of these points.

| | |
|---|---|
| *A*(5, 2) | *B*(8, 4) |
| *C*(5, 1) | *D*(5, −4) |
| *E*(3, −4) | *F*(3, −1) |
| *G*(−2, −1) | *H*(−2, −4) |
| *I*(−4, −4) | *J*(−4, 1) |
| *K*(−5, 3) | *L*(−6, 2) |
| *M*(−8, 2) | *N*(−8, 3) |
| *P*(−6, 5) | *Q*(−5, 5) |
| *R*(−5, 6) | *S*(−4, 5) |
| *T*(−2, 2) | *U*(5, 2) |

**b)** Join the points in alphabetical order, with straight lines. (Join *A* to *B*, *B* to *C*, *C* to *D*, … and so on. Finish by joining *U* to *A*.) What picture have you drawn?

**8 a)** On a piece of graph paper, draw and label the *x*- and *y*-axes to make a Cartesian plane. Now plot (and label) each of these points.

| | |
|---|---|
| *A*(−3, −6) | *B*(−2, −4) |
| *C*(−1, −2) | *D*(0, 0) |
| *E*(1, 2) | *F*(2, 4) |
| *G*(3, 6) | *H*(4, 8) |

**b)** Join the points in alphabetical order, with straight lines. Finish at *H*. What shape is the graph you have drawn?

**9 a)** On a piece of graph paper, draw and label the $x$- and $y$-axes to make a Cartesian plane. Now plot (and label) each of these points.

$A(-3, 5)$ $\qquad$ $B(-2, 0)$

$C(-1, -3)$ $\qquad$ $D(0, -4)$

$E(1, -3)$ $\qquad$ $F(2, 0)$

$G(3, 5)$

**b)** Join the points with as smooth a curve as you can.

Can you find out what we call this kind of curved graph? You will learn about them in Year 9.